D1165293

The Hutchinson Atlas of Battle Plans

THE HUTCHINSON ATLAS OF BATTLE PLANS

Before and After

Advisory Editor

Richard Holmes

General Editor

John Pimlott

Helicon

Helicon Publishing Ltd
42 Hythe Bridge Street
Oxford OX1 2EP
e-mail: admin©helicon.co.uk
Web site: http://www.helicon.co.uk

First published 1998

ISBN 1-85986-241-1

British Cataloguing in Publication Data

A catalogue for this book is available from the British Library

Typeset by TechType, Abingdon, Oxon
Printed and bound in Great Britain by
The Bath Press

Editorial

Clare Collinson
Katie Emblen
Alyson McGaw
Hilary McGlynn
Anne-Lucie Norton

INDEXER
Drusilla Calvert

Design and Production

PRODUCTION
Tony Ballsdon

ART AND DESIGN
Terence Caven

CARTOGRAPHY
Olive Pearson

Contents

Introduction

The 19th-century German military thinker Carl von Clausewitz has come in for his fair share of criticism as our own century spins to its close. But although his work may variously be repetitious, obvious, obdurate, and narrow-minded, there can be little doubt that it illuminates many of the ageless truths of war. Two of these are of particular importance in the pages that follow. 'War ...' declared Clausewitz, 'is not the action of the living force upon a lifeless mass (total non-resistance would be no war at all) but always the collision of the living forces.' 'The will', he warned, 'is directed at an animate object that *reacts*.' War is a mutual activity, whose theoretical scope is restricted by, amongst other things, the fact that one's opponent is also struggling to win.

Next, he went on to emphasize that war was an environment in which: 'Everything looks simple; the knowledge required does not look remarkable, the strategic options are so obvious that by comparison the simplest problem of higher mathematics has an impressive scientific dignity.' However, because of friction – 'the factors that distinguish real war from war on paper' – nothing is ever quite what it seems. Soldiers are often hungry, thirsty, tired, and frightened: their martial zeal is cor-roded by what Shakespeare's Henry V called 'rainy marching in the painful field'. Individuals and whole units get delayed or lost. Supplies of food and ammunition run short. Orders go astray, or are not understood even when they do arrive. Niall Barr's study of Waterloo (1815) and Andrew Lambert's chapter on Balaklava (1854) both show the impact of badly worded and misunderstood orders.

Friction clouds thought. Under the pressure of the moment sensible commanders have lapses of judgement. Frederick the Great may have been a towering genius, but at Kunersdorf (1759) poor reconnaissance meant, as Tim Bean recounts, that his attack had little chance of success: he tried to make the ground fit the plan, not vice versa. In such an unforgiving environment, clear, directing thought, widely understood, is at a premium. G D Sheffield suggests that at Gallipoli (1915) there were actually too many commanders: differences of opinion between the naval and military commanders, the fruit of a lack of common doctrine, got the campaign off to a bad start from which it never recovered.

And as Gallipoli demonstrated, terrain and climate conspire against combatants. Unexpected rain turns babbling brooks into impassable torrents, early snow blocks passes, repeated rain rusts soldiers' spirits. The enemy, as we have seen, is a player in his own right, and may well decide that, of the three courses open to him, he will adopt a fourth. Luck, what Frederick called 'His Sacred Majesty Chance', is hyperactive. In short, as Clausewitz observed: 'Everything in war is simple, but the simplest thing is difficult.'

The combined effects of the enemy's activities and of the friction which characterizes war are always a strain and sometimes a surprise. It is small wonder that commanders, across most of military history, have worked hard to limit their impact. Planning plays a key role in this respect, in striving to limit the impact of friction on one's own side and compound its effects upon the enemy.

This book is primarily concerned with battles, but it is important to remember that these do not occur in isolation. Late 20th-century western military thought identifies three levels of war, the strategic, the operational, and the tactical. At the highest level, strategy, the business of governments, involves all the state's agencies in pursuing its general good. Military strategy is concerned with the use, actual or threatened, of the military instrument within this broader context. Politicians and military officers recognize that war is, at the very least, as Clausewitz put it, a political act, 'merely the continuation of policy by other means'. The wiser heads amongst them go deeper, acknowledging that it is an instrument whose use is often inappropriate: many complex problems are simply not amenable to military solutions, and the use of force may well make bad situations worse.

Much of the logic of strategy is timeless, and rises above changes in technology and tactics. At the other end of the scale, tactics is concerned with fighting individual battles and engagements. Although there are several tactical ploys which have validity across time – as Ian Hogg tells us, the Confederate assault at Chancellorsville (1863) succeeded largely because it was a classic attack on the flank of a line, usually a weak point – tactics tends to be technology-related, and thus subject to continuous evolution. However, as many commanders across history have failed to appreciate, battles are rarely ends in themselves, but ought to be stepping-stones on the path to attaining the political goal for which the war is fought. Structuring battles and engagements so that they do indeed form a cohesive plan within a theatre of war is the realm of operational art, that level of war which links the tactical to the strategic. It is fair criticism of much of classical western military thought to suggest that it has, in the past, often been excessively preoccupied with battle. Military doctrines which have emphasized the avoidance of battle as part of a long drawn-out war designed to exhaust an opponent, or have shunned the head-on assault on an enemy's main strength in favour of indirect attack on a weak point had often much to commend them, especially if they reflected the cultural strengths and weaknesses of the combatants.

Failure to understand one's opponent ranks high amongst the causes of many a disaster. It is fairer to describe Isandlwana (1879) as a Zulu victory rather than merely a British defeat, and Edmund Yorke is right to pay tribute to the 'sheer raw courage, fighting spirit, and resilience of the Zulu army', as well as to emphasize fundamental British mistakes, notably dividing one's force before the main body of the enemy had been located. Tim Bean tells us how a ponderous Austro-Russian army made the fatal mistake of underestimating Napoleon, who was at his best at Austerlitz (1805), and Duncan Anderson's account of the fall of Singapore (1942) illustrates the folly of an overstretched defence policy shot through with disdain for the rising power of Japan.

It would, though, be a misreading of history to blame ancient, medieval, or early modern commanders for failure to grasp the importance of the operational level of war which is, after all, a relatively modern concept. The wars of their age were sometimes decided by a single major clash, especially if, as was often the case, the losing commander-in-chief was also head of state, and was removed from the scene at a stroke. Hastings (1066) struck Saxon England a blow from which it was never to recover, and Mohacs (1526) effectively blotted Hungary from the map for over a century. But, interestingly enough, battle was often indecisive even in the Middle Ages: a series of stunning English victories in the Hundred Years' War, one of them, Crécy (1346), brilliantly described by Matthew Bennett, could not prevent the eventual ruin of the English cause in France.

We would expect modern military plans to be powerfully influenced by the political aim of the war, and to be directed towards the achievement of a military endstate which political leaders would recognize as constituting success. In practice this success is more likely to be achieved by some sort of compromise peace than by the total overthrow of one of the contending parties. Clausewitz warned that this theoretical extreme of war, in which the victor totally destroyed his opponent's army, was likely to be much modified in the real world. However, Hannibal's victory over the Romans at Cannae (216 BC) – a classic example, as Sean McKnight tells us, of the tactical ploy of double envelopment – was so remarkable as to exercise a fatal attraction on some subsequent commanders, most notably Alfred von Schlieffen, chief author of the German plan for the invasion of France in 1914.

Modern military theorists tend to favour what we would term manoeuvre warfare, which seeks the psychological dislocation of the enemy's command structure rather than the simple destruction of his combat assets. In his examination of the German attack in the West in 1940, Sean McKnight rightly emphasizes 'the shock of surprise' which played such a great role in this stunning victory. Part of the reason for German success lay in the technical and tactical surprise generated by the combination of tank and dive-bomber, and Lloyd Clark's study of Cambrai (1917), the first battle in which tanks were used on a large scale, shows just how powerful technical surprise can be. Yet there are times when plans have embodied, not high-risk, high pay-off masterstrokes, but what Niall Barr, in his chapter on Alamein (1942) calls 'the proper application of overwhelming force'. Montgomery recognized that he was likely to win if he fought what he was fond of describing as 'a teed-up battle', while a more fluid scheme might dissipate his numerical advantage and enable the Germans to make the most of their tactical slickness.

The key words here, of course, are 'proper' and 'overwhelming'. The British plan for the First Day of the Somme (1916), was, as Lloyd Clark terms it, a 'great illusion'. Although there was resolute infantry aplenty, there were too few guns, especially heavy guns, for the frontage attacked, and British inability to destroy German deep dugouts and to reach out into the depth of the German position was a fatal flaw in a plan which was, as plans so often are, oversold to its participants. Many historians would now agree that, ghastly though it was, the Somme did achieve a favourable balance in attritional terms, and that the British army which emerged from it was far better trained than that which began it. But the fact remains that the plan for the first day was unrealistic, and that the best of a generation died proving it.

The question of planning for the Somme highlights the balance between rigidity and flexibility which planners seek to strike. An attack on a wide front by hundreds of thousands of men with substantial artillery support has implications which go well beyond tactics: the supply of food, water, and ammunition, the handling of prisoners of war, casualties, and refugees, and the replacement of worn-out units are amongst the issues which demand serious thought, and if planners fail to take them into account they are failing in their duty. Yet a plan which marches at the steady pace of logistic detail may neither be able to cap-italize on a fleeting opportunity nor react, in a timely fashion, to an unexpected threat.

The serious doubts in the minds of many British commanders in the run-up to the Somme were suppressed by a chain of command which emphasized compliance and warned that criticism would rebound on the heads of the critics. This sort of policy should sound a warning note, and I maintain that one of the attributes of truly great generalship is the ability to comprehend loyal opposition based on a genuine understanding of local circumstances while refusing to be deflected from the aim. The German offensive in the Ardennes (1944) is an even more extreme example of an unrealistic plan. Stephen Badsey's expression 'progressive unreality' has it in a nutshell. It was possible for the Germans to concentrate a powerful force opposite a weakly held sector, and to attack at a time when the weather had grounded Allied aircraft. But it was difficult to mint operational coinage from this tactical success, and impossible to turn it into anything of strategic value. Many senior German officers were uncomfortably aware of this, but the role of Hitler – at times commanding at all three levels of war – ensured that German forces were, yet again, called upon to execute an impossible mission.

Some of the plans described in the pages that follow worked well, usually because of a good balance between conception and execution which, between them, wrong-footed the opponent. Others failed because they were fundamentally unrealistic in the first place, or, like the Union plan for Chancellorsville, because they were rendered irrelevant by an enemy who got inside their time cycle to rap out new ventures at a faster tempo. Yet sometimes neither side had a real advantage in planning, technology, or weight of numbers, and the battle was decided by the rugged fighting qualities of the contending armies, with victory and defeat resting in the narrowest of margins. G D Sheffield shows us just how close the Union came to defeat at Gettysburg (1863) – had the 20th Maine not risen to the hour the result could easily have been different. The British succeeded in preventing a German breakthrough at the First Battle of Ypres (1914), as Stephen Badsey tells us, by the merest wafer of khaki. Sir John French, the British commander in chief, may have been given to hyperbole, but he spoke the truth when he said that his only reserves were the sentries at his gates. Finally, as Ian Hogg so graphically observes, the German airborne attack on Crete (1941) led to the defeat of both sides. The Germans captured the island, but such were the casualties suffered by their parachute arm that it never mounted a large-scale assault again.

It is tempting, sitting in comfortable surroundings with all the facts at one's disposal, to be critical of many of the plans which have unrolled across the battlefields of history. After all, their impact has been horrific at the human level, spreading what T E Lawrence called 'rings of sorrow' through communities blighted by bereavement. But one of the virtues of this book is to show just what a complicated business war really is, and just how few of its bets are safe. 'If no one had the right to give his views on military operations except when he is frozen, or faint from heat and thirst, or depressed from privation and fatigue, objective and accurate views would be even rarer than they are', noted Clausewitz. 'But they would at least be subjectively valid, for the speaker's experience would precisely determine his judgement.'

While some of history's senior commanders still emerge as impressive, even sympathetic, figures, others seem far less appealing, and from our perspective the self-confidence and bombast which often formed part of their stock in trade are more likely to irritate than inspire. Yet there are times when, dare I say, they deserve our compassion. By the time that they were called upon to lead great armies, many had become more worriers than warriors, weighed down by the burden of command, and well aware of the human consequences of their plans. Even Wellington's robust personality could not insulate him against personal grief. After Waterloo, he remarked that, barring a battle lost, nothing could be more painful than to win a battle with the loss of so many of one's friends.

This last phrase strikes a chord. I know that all contributors to this book will join me in lamenting the untimely death of a mutual friend. Dr

John Pimlott, its general editor, did not live to see the completion of a project into which he plunged with his characteristic energy. I hope that he would not be disappointed in its result.

RICHARD HOLMES

Notes on the Contributors

Advisory Editor

Professor Richard Holmes of Cranfield Security Studies Institute is the author of numerous books on the history of warfare including *Firing Line*, and a frequent television presenter, most recently for his celebrated *War Walks* series.

General Editor

Dr John Pimlott was until 1997 Head of the War Studies Department at the Royal Military Academy Sandhurst. He was the author or editor of numerous books on warfare, including *Armed Forces and Modern Counter-Insurgency, The Gulf War Assessed*, and *Vietnam: The Decisive Battles*.

Special Adviser and Contributor

Ian Hogg served in the British Army for over thirty years, and is an authority on all aspects of artillery and small arms. He is the author of over a hundred books on military history, and editor of *Jane's Infantry Weapons*.

Contributors

Dr Duncan Anderson MA DPhil (Oxford) is head of the War Studies Department at RMA Sandhurst. He has written extensively on the Second World War in the Far East and the Pacific. His latest work, a study of the Battle for Manilla in 1945 was coauthored by Col Richard Connaughton and the late Dr John Pimlott.

Dr Stephen Badsey MA (Cantab) FRHistS is a Senior Lecturer at RMA Sandhurst and a Senior Research Fellow of De Montfort University, Bedford. He is a specialist on military theory and on media presentations of warfare.

Dr Niall Barr is a Senior Lecturer at RMA Sandhurst. He has led military battlefield tours of Alamein, and is currently researching a book on Montgomery and the Eighth Army.

Tim Bean is a Senior Lecturer at RMA Sandhurst, and a specialist on 18th-century warfare and naval history.

Matthew Bennett MA FSA FRHistS is a Senior Lecturer at RMA Sandhurst. He is an expert on medieval warfare, and author of many works on the subject including *The Cambridge Atlas of Warfare: The Middle Ages 787–1485*.

Lloyd Clark is a Senior Lecturer at RMA Sandhurst, and has published widely on 20th-century conflict. A keen student of World War I, he is currently writing a book about the development of British fighting methods 1914–18.

Dr Andrew Lambert FRHistS is a Senior Lecturer in the War Studies Department at King's College, University of London. He is a specialist on both naval warfare and the Crimean War, and author of numerous works including *The Crimean War* and *The Last Sailing Battlefleet*.

Sean McKnight is Deputy Head of the War Studies Department at RMA Sandhurst and a Senior Research Fellow of De Montfort University, Bedford. He is a specialist on warfare in the Middle East, from medieval times to the present day, and is currently researching on the Mesopotamia campaign in World War I.

Dr G D Sheffield is a Senior Lecturer at RMA Sandhurst, a Senior Research Fellow of De Montfort University, Bedford, and a former Secretary-General of the British Commission for Military History. He has written widely on the British Army in World Wars I and II, and has most recently edited *Leadership and Command, The Anglo-American Experience Since 1861*.

Dr Edmund Yorke is a Senior Lecturer at RMA Sandhurst. He has published widely in Defence, Commonwealth, and African affairs. His qualifications include a BA in Modern History and International Relations from Reading University, an MA in British Commonwealth History from London University, and a PhD in African History from Cambridge University.

General Map Keys

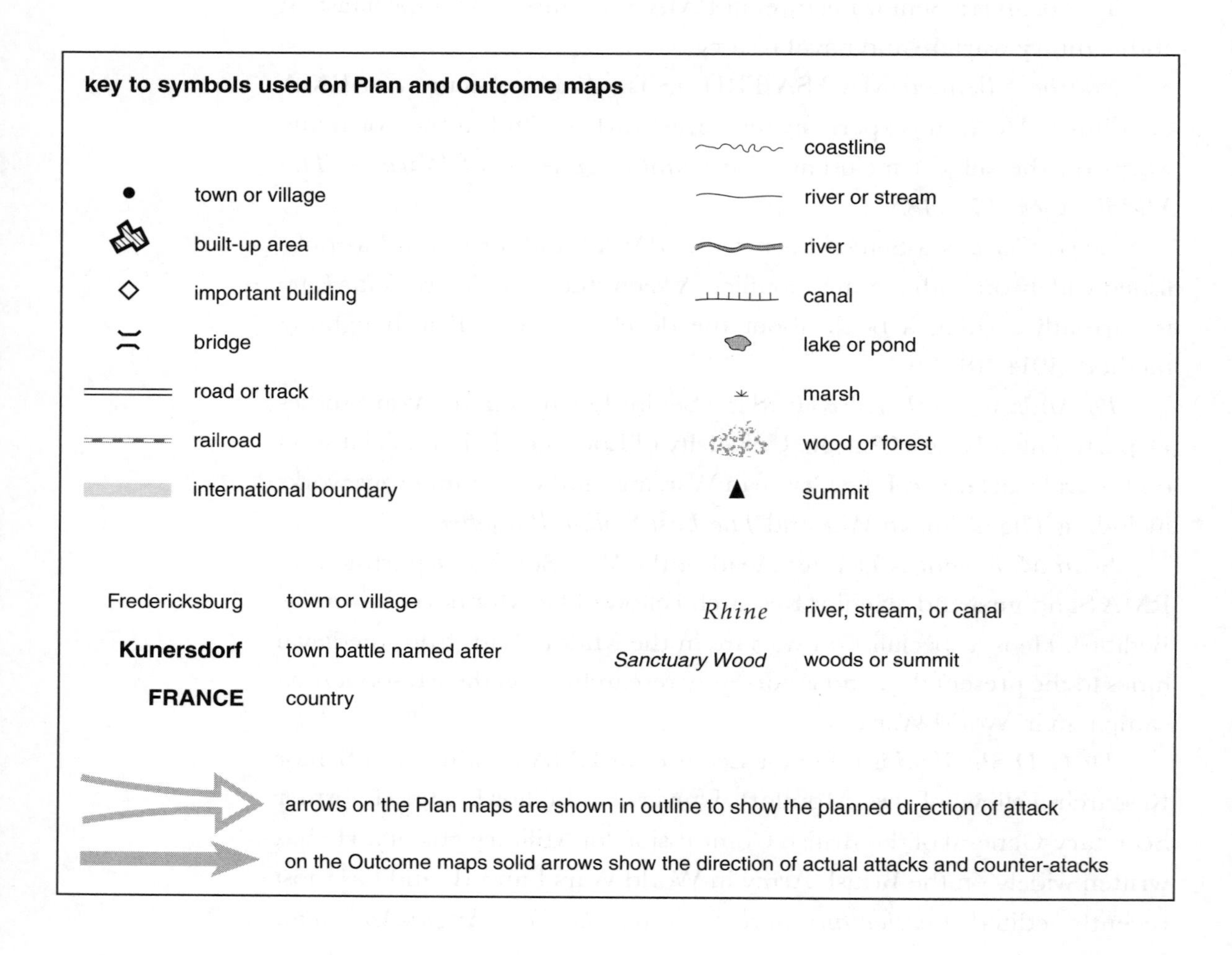

key to symbols used on locator maps

Ypres location of battle

present day international boundary

The Classic Ploys

'Nothing succeeds in war except in consequence of a well-prepared plan.' Napoleon

The army or leader that always follows the same plan for every battle will eventually be defeated. No matter how good a plan might be, an enemy will eventually come to expect it, and work out a way to counter it. Even so, some of the most simple and effective battle plans have throughout history been rightly regarded as the 'classics', plans which if they can be made to work will produce decisive victory for the side which uses them. These plans have influenced the way battles have been fought for centuries, they have been adopted and used in a wide variety of circumstances, and they have always been among the favourites of the great commanders. The three battles which follow illustrate three of the most effective and widely used of these classic ploys.

Cannae

2 August 216 BC

The Double Envelopment

'As for the Romans, after this defeat they gave up all hope of maintaining their supremacy over the Italians, and began to fear for their native soil, and indeed for their very existence ...' Polybius

Beyond doubt the deadliest of the classic ploys of battle is the double envelopment, which aims not simply to defeat the enemy forces but to surround them and crush them out of existence. This is the most difficult of all the manoeuvres of warfare to execute successfully, since it very often requires the unwitting cooperation of the enemy in their own destruction by obligingly continuing to advance into a trap. It requires greater powers and speed of manoeuvre, better quality troops, and great confidence on the part of any commander who intends to use it. But the rewards are equal to the risks if it succeeds.

Extraordinary as it may seem, much of modern thinking on the double envelopment manoeuvre still comes from a battle fought more than two thousand years ago. On 2 August, in the year 216 BC, the armies of Rome and Carthage fought the largest battle of the Second Punic War (218–201 BC) near the town of Cannae in southern Italy. The battle was a crushing defeat for the forces of Rome, despite their substantial numerical advantage, and their formidable heavy infantry. The Carthaginian success was due above all to the brilliance of the battle plan formulated by their commander Hannibal, and his skill in executing this plan.

The Strategic Background

Although the Romans had won the First Punic War (264–241 BC) against Carthage, it had not been a decisive victory. Carthage recovered its prosperity and added to its empire (modern Tunisia and the coastal plains westward towards Morocco) by conquering southern Spain. This conquest was the project of the Carthaginian leader Hamilcar Barca, who intended to use Spanish resources in order to resume the war with Rome. In 221 BC Hamilcar's son Hannibal became commander of the Carthaginian forces in Spain, and it fell on his shoulders to attempt to reverse the verdict of the First Punic War.

Hannibal recognized that control of Italy gave Rome access to a military manpower that far exceeded that of Carthage. He planned to invade Italy not in order to destroy the city of Rome, but to encourage Rome's allies to defect, so depriving Rome of much of its military strength. Rome's superior strength at sea forced Hannibal to invade the Italian peninsula by land from Spain. The army Hannibal commanded had few Carthaginians in its ranks, as Carthage fought its wars with mercenaries and levies from its empire. When it reached northern Italy in the winter of 218 BC it consisted of 20,000 infantry (8,000 Spanish and 12,000 African) and 6,000 cavalry (Numidian and African). Hannibal immediately started recruiting from the Gallic tribes of northern Italy, who had only recently been conquered by the Romans. In December Hannibal defeated a Roman army of 40,000 on the River Trebia, and this victory helped him consolidate his alliance with the Gauls.

In contrast to the Carthaginians, the Roman Republic drew upon its own citizens for its military manpower. Indeed, except in dire emergencies, Rome raised its armies from citizens wealthy enough to provide their own equipment, and service was seen as a privilege and a duty. Roman forces also included separate allied contingents (Latin, Italic, Greek, and even Gallic) provided by Italian states subordinated to Rome. The Greek historian Polybius estimated Rome's potential military manpower as 273,000 Romans (excluding those too poor to serve), 85,000 Latin allies, and 276,000 Italian allies.

Command of the Carthaginian army in Spain was a Barca family affair: upon Hamilcar's death, his son-in-law Hasdrubal inherited the command, and Hannibal succeeded Hasdrubal when he was assassinated in 221 BC. Carthaginian war leaders often enjoyed prolonged periods in command, and under Barcid leadership the army in Spain had an impressive group of experienced commanders. In Italy Hannibal's army of necessity enjoyed great continuity in command, and the quality of Hannibal's leadership gave Carthaginian forces in Italy a considerable advantage over their Roman opponents. Roman armies were normally commanded by annually elected 'praetors' and 'consuls' (the two consuls were the senior magistrates of the Republic). This could produce incompetent commanders, and continuity of command was often disrupted as magistrates were forbidden from standing for immediate re-election, though it was possible to have a command extended. These disadvantages should not be exaggerated: it is only in comparison to Hannibal that his Roman opponents seem so inadequate. Roman nobles were rarely military novices, and most acquired military experience as junior officers, often by being elected *tribuni militum* (tribunes of the soldiers) early in their careers.

The Invasion of Italy

In 217 BC Hannibal moved south into Italy, and in June he won the Battle of Lake Trasimene, destroying a Roman army of 25,000 and killing its commander, the consul Flaminius. Defeat precipitated a political crisis in Rome, and Q Fabius Maximus was elected to the unusual temporary post of 'dictator'. Fabius was convinced of Hannibal's tactical genius, and decided to avoid confronting him in any pitched battles, preferring instead to wear him down with a strat-egy of small engagements (the term 'Fabian strategy' comes from Fabius' methods) and a scorched-earth policy. Despite Hannibal's military successes, he had not attracted any Italic or Latin city-state to defect from Rome. But victory at Trasimene enabled him to move south of Rome, where he hoped to encourage defectors.

Fabius' policy of delay stemmed the tide of Carthaginian victories, but proved increasingly unpopular in Rome, especially when Hannibal started to destroy nobles' estates in a methodical manner. In early spring 216 BC, Fabius' faction lost out in the consular elections, and both of the new consuls (Caius Terrentius Varro and Lucius Aemilius Paullus) favoured an offensive strategy, aiming to crush Hannibal with overwhelming numbers.

Traditionally a consular army consisted of two 'legions' (the basic Roman military unit of about 5,000 men) and a similar number of allied troops. But for the campaign of 216 BC each consul received four strengthened legions, and their commands were combined. The two consuls thus had joint command of an army with 80,000 infantry and 6,000 cavalry, roughly half of which were allied forces.

Hannibal also sought a large set-piece battle, as he urgently required an impressive military victory to entice Rome's allies to change sides. In June he moved his army rapidly into the region of Apulia, seizing the Roman supply base of Cannae, which lay a few miles from the sea on the River Aufidus. This secured sufficient food for Hannibal's army and cut the Romans off from their normal source of supply. It also placed the large consular army, which had just arrived, in an awkward position: it would either have to retreat, cope with a very disadvantageous supply situation, or attempt to engage Hannibal in battle. However, the terrain near Cannae was relatively flat which favoured cavalry, the arm in which the Carthaginians enjoyed both a qualitative and quantitative advantage.

As both consuls were with the army, they elected to command on alternate days. Tradition paints Varro as hot headed, and his colleague Aemilius as wisely cautious. But this traditional view is very doubtful, as both consuls made haste to confront the Carthaginian army. At dusk on 30 July, a day under Aemilius' command, they came up with Hannibal's army, and Aemilius took the aggressive step of sending a strong force over the river to build a fortified camp and harass the Carthaginian foragers.

For the next two days, the Romans occupied two camps astride the River Aufidus. A few miles upstream, the Carthaginian army was initially concentrated on the right bank of the river, near Cannae. However, as the bulk of the Roman force remained on the left bank, Hannibal ordered his army to cross the river and established a new camp. This cut the Romans off from broken ground to the southwest, while the terrain on the left bank, which does not rise above the 50 foot contour for six miles to the coast, favoured Hannibal's cavalry. Neither consul liked this ground, and on both 31 July and 1 August the Romans refused battle despite being harassed by Hannibal's Numidian horse.

The Rival Plans

The Roman army was now in an untenable position, and it would either have to withdraw or accept the risks of battle. Just after dawn on 2 August Varro ordered the army to deploy for battle on the right bank, where the ground rises away from the Aufidus and a ridge runs alongside the river. Varro sensibly elected to fight on ground which was flat enough for his infantry to deploy, but which although it was reasonable terrain for cavalry was not ideal, especially near the river. The Roman battle line faced broadly south and the Carthaginians north, so neither army was inconvenienced by the sun, but it seems that the local wind, the *Volturnus*, carried clouds of dust into the Romans' faces, obscuring their view of the battlefield.

Varro's army was drawn up in an orthodox manner for the Romans, with cavalry on both flanks and infantry in the centre. There were roughly equal numbers of Roman and allied infantry, a total of 55,000 men, and blocks of allied infantry were interspersed with Roman troops. Aemilius commanded the right wing, which consisted of the 2,400 Roman cavalry and 10,000 infantry. The centre of the Roman line, with 25,000 infantry, was commanded by Cn Servilius Geminus and M Minucius Rufus. On the left, Varro took personal command of 20,000 infantry and 3,600 allied cavalry. This made a total of 55,000 infantry and 6,000 cavalry. In front of the main body of the army were roughly 15,000 lightly armed skirmishers. A further 10,000 infantry, probably drawn from Aemilius' consular command, were left guarding the camps.

The skirmishers in front of the Roman army, known as *velites*, were the youngest and poorest recruits, equipped with javelins and swords, but with no body armour; *velites* fought enemy skirmishers, disrupted the main body of the enemy, and then retired behind the line infantry. Behind the *velites* were the heavy infantry of the line, equipped with the *scutum* (a large oval shield) and body armour, although since the recruits provided their own equipment there was considerable variation in the quality of the armour. Legions stood in blocks divided into three ranks of infantry: the first two ranks were further divided into ten 'maniples', each composed of two 'centuries' (with about 80 men in a century); the last rank consisted of ten mani-

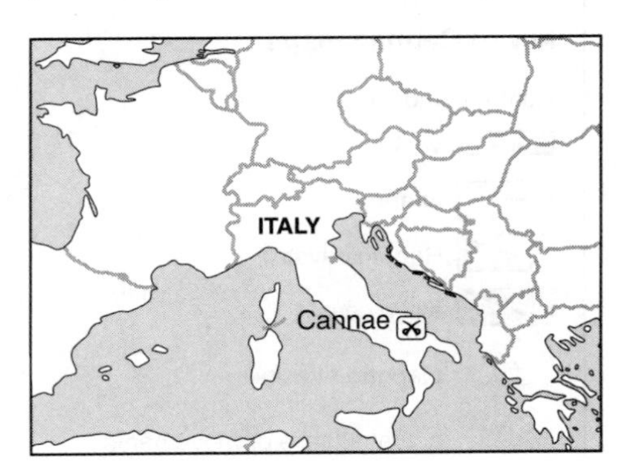

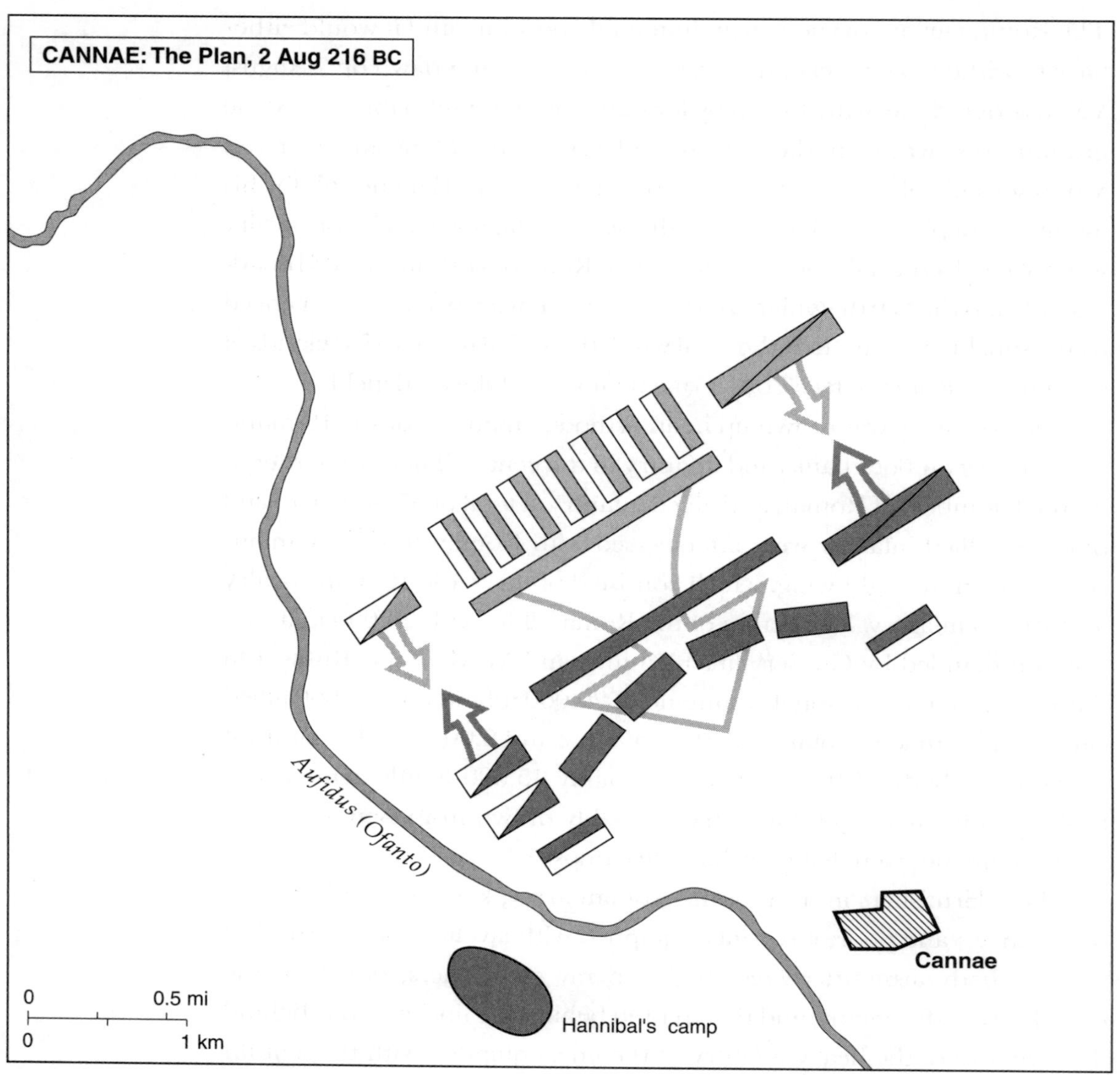
CANNAE: The Plan, 2 Aug 216 BC
Aufidus (Ofanto)
Cannae
Hannibal's camp
0
0.5 mi
0
1 km

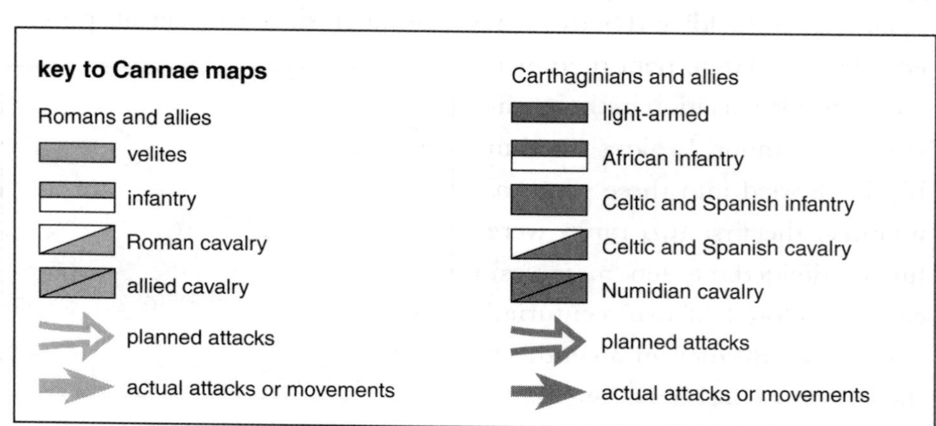
key to Cannae maps
Romans and allies
velites
infantry
Roman cavalry
allied cavalry
planned attacks
actual attacks or movements
Carthaginians and allies
light-armed
African infantry
Celtic and Spanish infantry
Celtic and Spanish cavalry
Numidian cavalry
planned attacks
actual attacks or movements

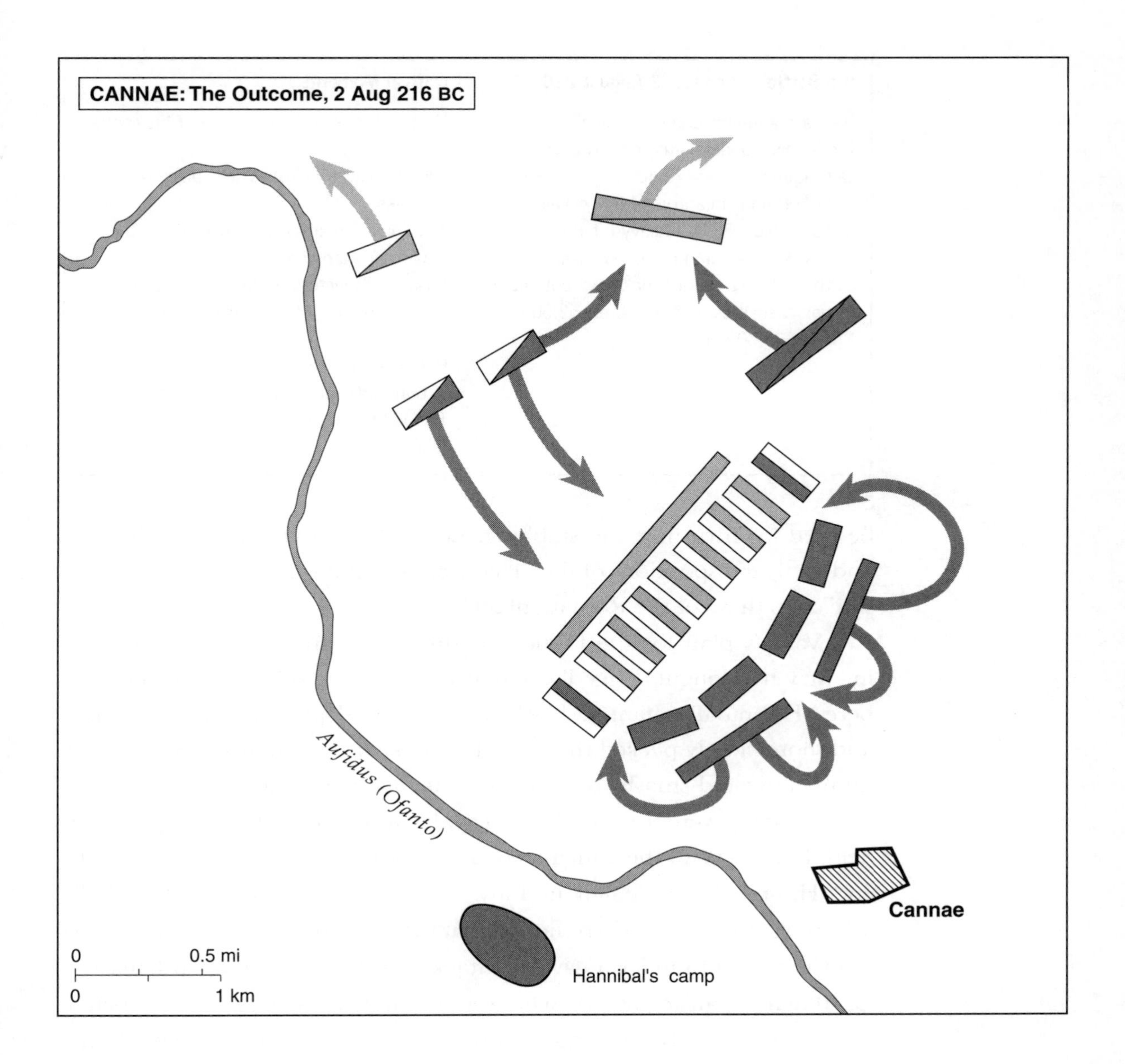

ples each consisting of a single century. The first rank were known as the *hastati*, armed with two javelins and short sword; behind them were the similarly armed *principes*; and bringing up the rear were the *triarii* who probably carried thrusting spears rather than javelins. The rank in which a Roman soldier served depended upon his experience and wealth, with the *triarii* being the best equipped and most experienced. The main weapon of Roman line infantry was the *gladius*, a short sword that could

The Battle of Cannae, 2 August 216 BC	Critical Moments
The Carthaginians under Hannibal The Romans under Caius Terrentius Varro Carthaginian forces: 28,500 heavy infantry, 11,500 light infantry, 10,000 cavalry Roman forces: 55,000 heavy infantry, 15,000 light infantry, 6,000 cavalry Carthaginian casualties: unknown but slight Roman casualties: approximately 52,000 dead, 19,300 prisoners	Hannibal deploys the infantry of his centre in a bulge The first cavalry clashes take place on both flanks The Roman infantry gets sucked into the Carthaginian centre Hasdrubal's victorious cavalry manoeuvres onto the rear of the remaining Roman cavalry Hannibal's African infantry swings in on both sides of the Roman infantry Trapped in the crush, the Roman infantry is annihilated

be used for thrusting and stabbing. Less is known about the equipment and tactical dispositions of the allied troops, but it is likely that these were similar to those of the Roman infantry.

Varro's plan for the battle was simple. In previous battles Roman infantry had fought doggedly, and at Cannae they substantially outnumbered Hannibal's infantry. The maniples were deployed in greater depth and more closely packed than usual. Varro expected that sheer weight of numbers would smash through the Carthaginian centre, and that once their infantry were destroyed their cavalry would almost certainly flee the field. In any event, the anticipated destruction of the enemy centre would end Hannibal's campaign in Italy. Although he respected Hannibal's advantage in cavalry, Varro hoped that the ground on the flank nearest to the river would hinder a cavalry attack, and that on the open flank the Carthaginian allied cavalry, who were numerically stronger and of better quality than the Roman horse, would hold long enough to enable his infantry to win the battle.

The Carthaginians also deployed with their stronger infantry in the centre, their cavalry on both wings, and approximately 11,500 skirmishing infantry out in front of the main body of the army. Nearest the river, Hannibal deployed 6,500 Gallic and Spanish cavalry commanded by another Carthaginian known as Hasdrubal. In the centre Hannibal and his brother Mago took personal command of roughly 28,500 infantry, of which 10,000 were Africans, 6,000 Spaniards, and 12,500 Gauls. On the open right flank were 3,500 Numidian cavalry commanded by Hanno.

There was much more variety in the Carthaginian force than in the Roman. The Gauls, who constituted the least reliable element of Hannibal's army, were armed with long slashing swords and fought almost naked. The Numidian cavalry was composed of highly skilled horsemen, who fought in loose order and carried spears for stabbing and hurling. The Spanish fought with a short stabbing sword, their infantry being more lightly armoured than their Roman enemies. Hannibal's best troops were his African infantry, whose main weapon was the sword, and who were equipped with looted Roman arms. Unlike the Roman *velites*, Carthaginian skirmishers seem to have been high-quality troops, including specialist infantry such as slingers from the Balearic Islands. Carthaginian skirmishers also probably retained thrusting spears and joined with the infantry of the line. Unlike the Romans, many of whom were new recruits, Hannibal's army was experienced, and with several victories under their belts, the troops' confidence in themselves and their commanders was high.

Hannibal's deployment of his cavalry seems perverse, as the most numerous force was placed next to the river, where space was constricted, and the ridge alongside the river further hindered movement. However, the Gauls and Spaniards were trained to fight in closer order, and were equipped for close combat on horseback or on foot. On the open flank of the Carthaginian left, the Numidians had the space to exploit their superior horsemanship. Hannibal's masterstroke was in his deployment of his centre, initially in a conventional straight line with about half the depth of the Roman formations opposite. Before the battle commenced, Hannibal moved the Gaulish and Spanish infantry in the middle forward, and hence the Carthaginian infantry line made a bow-shaped formation with the African infantry standing further back on either side. This novel infantry formation was designed to suck the Roman infantry into the centre, first taking the impetus out of their attack and then, as the bulge in the Carthaginian line was pushed back, shortening the line and so providing a stiffer resistance. The African infantry on either side of the bulge had the key role of attacking both sides of the advancing Roman infantry, thereby constricting them into such a small area that they could not use their numbers effectively. If the Carthaginian cavalry, having disposed of the enemy horse, were able to join the main battle this would complete the envelopment of the Roman army. Hannibal planned to do more than win the battle; he planned to annihilate the Roman army.

The Battle and its Aftermath

The battle opened with both sides' skirmishers clashing inconclusively. The *velites* then played no further active role in the battle, but it is not known where the Carthaginian skirmish line went and it is a reasonable assumption they continued to participate actively. The Carthaginian horse nearest the river charged, but lack of space and the terrain forced a rather static engagement in which some of the cavalrymen on both sides fought dismounted.

Just as the Roman cavalry were giving way, the main infantry battle commenced, and although the badly outnumbered Spanish and Gallic infantry fought bravely they were pressed back. The Roman advance was initially successful, sucking them into the Carthaginian centre. But while the Romans pressed their advantage against the bulge of the Carthaginian infantry, the resistance of the Roman cavalry was ending. The fighting by the river had been vicious, and most of the Roman cavalry had been killed. Often in ancient battles the victorious cavalry on one wing would pursue its defeated enemy off the battlefield, so playing no further role in the fighting, but the pedestrian pace of Hasdrubal's advance made it easy for him to keep his force in hand. He moved his cavalry behind the Roman infantry, so threatening the Roman allied cavalry on the open flank who up to that point had been inconclusively engaged with the Numidians. Meanwhile, the African infantry attacked simultaneously on both flanks of the Roman infantry, in a fairly complex manoeuvre which suggests considerable tactical flexibility and well-trained troops.

Now the battle was going badly wrong for the Romans, as the sudden appearance of Hasdrubal's cavalry almost in their rear precipitated the flight of the largely intact Roman allied cavalry, hotly pursued by the Numidians. In the centre the African attack from both sides caused chaos in the Roman infantry formations as they attempted to turn to face the new threat. The Roman assault on the Gallic and Spanish infantry stalled, and the legions found themselves constricted into such a small space that only a small proportion of their infantry could even bring their arms to bear. Finally, the attack of Hasdrubal's cavalry into their rear sealed the Romans' fate, converting disaster into catastrophe. The trapped Roman infantry were slaughtered, although Polybius' claim that roughly 70,000 were killed is almost certainly exaggerated, and the Roman historian

Livy's figure of 52,000 Roman dead (47,500 infantry and 2,700 cavalry) seems more realistic. According to Livy the Carthaginians also took 4,500 prisoners on the battlefield, and a further 14,800 in the aftermath of the battle, including those guarding the two camps. Of the great Roman army which had begun the battle, only 14,000 troops survived it.

This was a dramatic victory, and Hannibal was urged by one of his commanders to march on Rome. He declined, and was told that he did not know how to use a victory. Many modern commentators have echoed this criticism, but this betrays an ignorance of Hannibal's strat-egy and the circumstances that he faced. It was 250 miles from Cannae to Rome, and the immense Roman reserves of manpower ensured that the city would be defended. In the aftermath of Cannae, the Roman Senate committed itself to continue the war, and elected a new dictator, M Iunius Pera, who immediately armed 8,000 freed slaves and 6,000 criminals to provide a garrison for the city. Ancient siege warfare was normally a very slow process, and Hannibal only kept his army intact in Italy by keeping it mobile. He believed that detaching Rome from its allies was the only way of permanently reducing its military strength, and a failed advance on Rome would detract from the impact that he hoped Cannae would have on these allies.

The great Carthaginian victory at Cannae did stimulate some important defections from Rome, including many Apulian communities in southern Italy, virtually all of Samnium, which made common cause with Hannibal, and the major city of Capua, which repudiated its Roman alliance. The victory also brought the Kingdom of Macedonia (in northern Greece) into the war against Rome. However, none of Rome's Latin allies deserted it, and most of its Italic allies stayed loyal, ensuring that its military potential remained great. Cannae helped Hannibal maintain his army in Italy for a further 13 years, but this most spectacular of victories was only a temporary setback in Rome's long campaign to crush its Carthaginian rival.

The victory at Cannae, in which the successful double envelopment meant that that enemy was not only defeated but destroyed, has inspired many modern military commanders. Even in the 20th century, the German Army's Great General Staff extolled the virtues of the 'Cannae manoeuvre' in both World Wars, and the United Nations coalition commander in the Persian Gulf War of 1991, the American General Norman

Schwarzkopf, also expressed his belief in the value of studying Cannae. Certainly, some of the most spectacular victories of land warfare owe much to the idea of annihilating an enemy by double envelopment, and Hannibal's defeat of a capable and numerically superior enemy was a tribute to his own brilliant generalship.

SEAN MCKNIGHT

Further Reading

De Beer, G., *Hannibal: The Struggle for Power in the Mediterranean* (London, 1978)

Delbrück, H., *A History of the Art of War* (London, 1975)

Lazenby, J. F., *Hannibal's War* (London, 1978)

Livy, *The War With Hannibal* (London, 1965)

Polybius, *The Rise of the Roman Empire* (London, 1979)

Chancellorsville

1–5 May 1863

THE FLANK ATTACK

'Our movement was a great success; I think the most successful military movement of my life.' Stonewall Jackson

To attack an enemy army frontally is to attack it where it is strongest. If it is forced to retreat, then it is pushed back onto its reserves and supplies, and may grow stronger still. Consequently, throughout history commanders have sought to avoid such frontal attacks by swinging their forces round to attack the enemy from the side and the rear – in military terminology, from the flank. So many advantages come from the flank attack that it becomes almost second nature in military planning, and of all the classic ploys it is the most obvious and most commonly attempted. But in trying to manoeuvre onto the enemy flank, an army inevitably exposes its own flank to the danger of a similar attack from the enemy. The risk is even greater if, in order to bring about the flank attack, it is necessary for a commander to split his forces into two or more parts in the presence of the enemy. This is well illustrated by a battle in the American Civil War, in which both sides had the same idea but one was much better led.

The Background to the Battle

By 1863 the Civil War had already lasted for two inconclusive years. In the east, strategy revolved around operations by both sides to try to capture the enemy's capital city: Union advances against the Confederate capital of Richmond in Virginia were matched by Confederate threats to

Washington. After the disastrous Union defeat at the Battle of Fredericksburg in December 1862, Major General Ambrose Burnside was replaced as commander of the main Union army in the theatre, the Army of the Potomac (named after the river which marked its general area of operations), by Major General Joseph E ('Fighting Joe') Hooker. Hooker's first task was to rebuild the morale of his army. He did this to such good effect that by April he was referring to it as 'the finest army on the planet' and making involved plans for an advance on Richmond. While this seemed a simple enough aim, it was made more difficult by the need to move so as to cover the approaches to Washington against the main Confederate army in the theatre, the Army of Northern Virginia under General Robert E Lee. Hooker's army lay on the River Rappahannock at Fredericksburg, which was almost due north of Richmond. But if Hooker took the obvious route south from Fredericksburg to Richmond, so outflanking Lee's army on the right, Lee and his Confederates might well slip to the left of Hooker's own flank and march on Washington. So the two sides would each capture the other's capital in a manoeuvre spoken of at the time as being like 'swapping queens' on a chessboard.

Lee was well aware of this dilemma, and had assessed the Union's options. The basic Union need was to get out of their position on the Rappahannock so as to obtain room to manoeuvre. The only way to achieve this would be for Hooker to move west and try to outflank Lee. An eastern flanking movement was, at that time of the year, impossible owing to swamps and widening of the river.

But before Hooker moved, there was another threat for Lee to consider, when the Union IX Corps landed from the sea at Newport News in Virginia, south of Richmond along the estuary of the River James. Lee speculated that the Union aim could be to attack North Carolina in support of operations inland, or to advance on the south side of the James so as to threaten Richmond. To try to contain this threat, Lee dispatched Lieutenant General James ('Old Pete') Longstreet with two divisions (those of Major General John B Hood and Brigadier General George E Pickett) southwards to resist any Union advance, while remaining close to the railroad near Richmond so as to be able to move back quickly if recalled. In addition, Longstreet was to send out foraging parties to build up the army's stock of food, and generally to supervise the Confederate defensive positions from the James to North Carolina. This was a mistake on Lee's part: Longstreet was jealous of the recent successes by Lee's other chief subordinate commander, Lieutenant General Thomas ('Stonewall') Jackson, and was also anxious to make a name for himself as an independent commander. With 40,000 troops under his command, Longstreet saw his opportunity for fame, and he pestered Lee with demands for even more troops so that he could take the offensive against the Union forces, even suggesting that Lee should retreat to a defensive position so as to yield up troops for his own planned attack.

As usual, Lee replied in a gentle and noncommittal manner, which did nothing to discourage Longstreet. At the end of March, when IX Corps re-embarked to be sent west, Lee suggested that it was about time that Longstreet came back and rejoined the main army. It is generally assumed that Lee's lack of authority here stemmed from a debilitating illness from which he was suffering at the time. Plagued by pains in his chest, back, and arms, and with a throat infection, Lee simply did not have the energy to act decisively with Longstreet, who now set out on a futile and pointless siege of the town of Suffolk, Virginia, held by Union forces. This meant that Lee's Army of Northern Virginia was seriously understrength when compared with the Army of the Potomac facing it on the north bank of the Rappahannock.

The Union Flank March

Lee's error in allowing Longstreet his head became apparent on 27 April when Hooker finally moved. He had with him a total of 138,378 men under arms with which to attack Lee's force of 62,000. With his forces already in motion, Hooker began his attack by launching 40,000 men under Major General ('Uncle') John Sedgwick straight across the Rappahannock just east of Fredericksburg on 29 April, in the same manner as Burnside before him. Lee considered this, conferred with Stonewall Jackson and his other commanders, and decided that this attack was a feint, and that Hooker would make his main attack some distance to the west. Leaving a single corps from his army to contain the Union attack, Lee began to move the rest of his forces to the west, trying to discern where Hooker would make his attack. At the same time he wired to Richmond that all available forces should be sent to him immediately, and a telegraph message was forwarded to Longstreet directing him to return to the Army of Northern Virginia as soon as possible. Three days later, Longstreet wrote back, 'I cannot move unless the entire force is moved... it would take several days to reach Fredericksburg... I will endeavour to move as soon as possible.' Lee was going to fight Hooker with a quarter of his army out of reach and miles from the battle.

Lee's cavalry division under its dashing leader Major General J E B ('Jeb') Stuart soon informed him that Hooker had crossed the Rappahannock, and then had continued south to cross its major tributary the Rapidan. Hooker then continued his advance southeasterly, so that Lee was now caught between two Union forces, with Sedgwick to his right and Hooker to his left. On the morning of 30 April the advance guards of Lee's and Hooker's forces met at a clearing called Chancellorsville (or Chancellor's Court House), on the edge of the heavily wooded area known locally as 'The Wilderness'.

The Rival Plans

While Hooker established his headquarters at the Chancellor family house, Lee's leading troops under Major General Richard H Anderson fell back and prepared a defensive line. Lee meanwhile rushed the major part of his army to join them. Lee had two advantages: first, Hooker had sent most of his cavalry off on destructive raids, so that he had only one brigade left to act as a reconnaissance force, while Lee had Jeb Stuart with his 5,000 horsemen as his eyes and ears; and secondly, the Wilderness area was so densely overgrown that it would prevent observation and effective fire from the Union artillery, a factor overlooked by Hooker. Lee therefore planned to attack Hooker and, having made sure that Sedgwick's activity around Fredericksburg was only a diversion, rode off to Chancellors-ville to find that a strange situation had developed.

Hooker had moved his forces forward from the Wilderness, largely so that his artillery could have open fields of fire. His forces were poised to sweep forward and overwhelm Lee, and quite probably would have done so. And then Hooker's nerve failed. Without any warning he suddenly ordered an immediate withdrawal from the open, to take up defensive positions within the protection of the Wilderness woods, claiming that he was now in a position to make Lee fight him on his own choice of ground. No convincing explanation for this move has ever come to light, although it is believed that Hooker was unnerved by intelligence reports which assumed that he had the full Army of Northern Virginia in front of him. Lee had taken the offensive, which as Hooker saw it he would be unlikely to do if he had fewer troops than the Union forces opposing him.

Hooker's generals could scarcely believe their ears. They argued, but to no avail, and like good soldiers they had to obey. But the move took the spirit out of the entire Union army: even the humblest soldier could see that they had been ready to end the war at a stroke, and now they were falling back.

Lee, also, could scarcely believe his ears when he heard that four Union corps had retreated in the face of five Confederate divisions. He suspected a trap or some devious manoeuvre. After analysing the situation and realizing what was happening, he rapidly reconnoitred the Union positions, finding that the enemy on the right had wisely pulled tight against the Rappahannock and fortified their positions with trees and obstacles. After a quick evening conference with Jackson, Lee sent two engineer officers off to study Hooker's other flank, and they soon returned to announce that the enemy line ended in thin air in the Wilderness without being anchored by any specific physical feature.

Once armed with this knowledge, Lee soon saw his opportunity: he would send Stonewall Jackson south on a convenient road, then west and north until he had outflanked the Union position completely. Jackson could then fall upon Hooker's rear from the northwest, the march being protected by a screen of Stuart's cavalry. When appraised of the plan, Jackson fell in with it at once, and outdid Lee in audacity by proposing to take his whole corps, leaving Lee with no more than a couple of understrength divisions to face the main part of Hooker's army. This also meant that Lee would have to take personal command of the troops facing Hooker, since his divisional generals were not competent to cope with such a situation without Lee's guiding hand. Nevertheless he agreed, and at 4.00 a.m. on 2 May Jackson began his move.

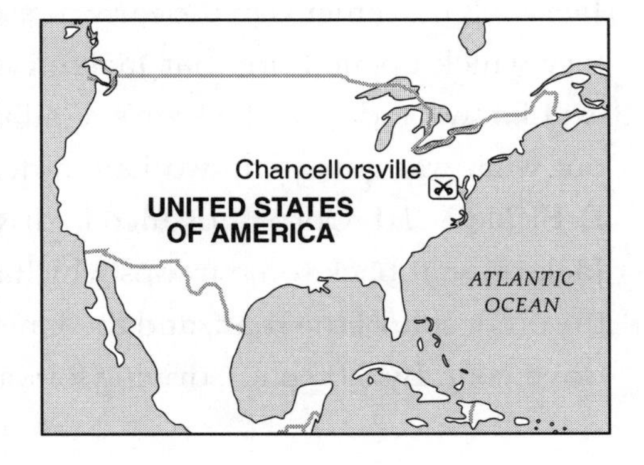

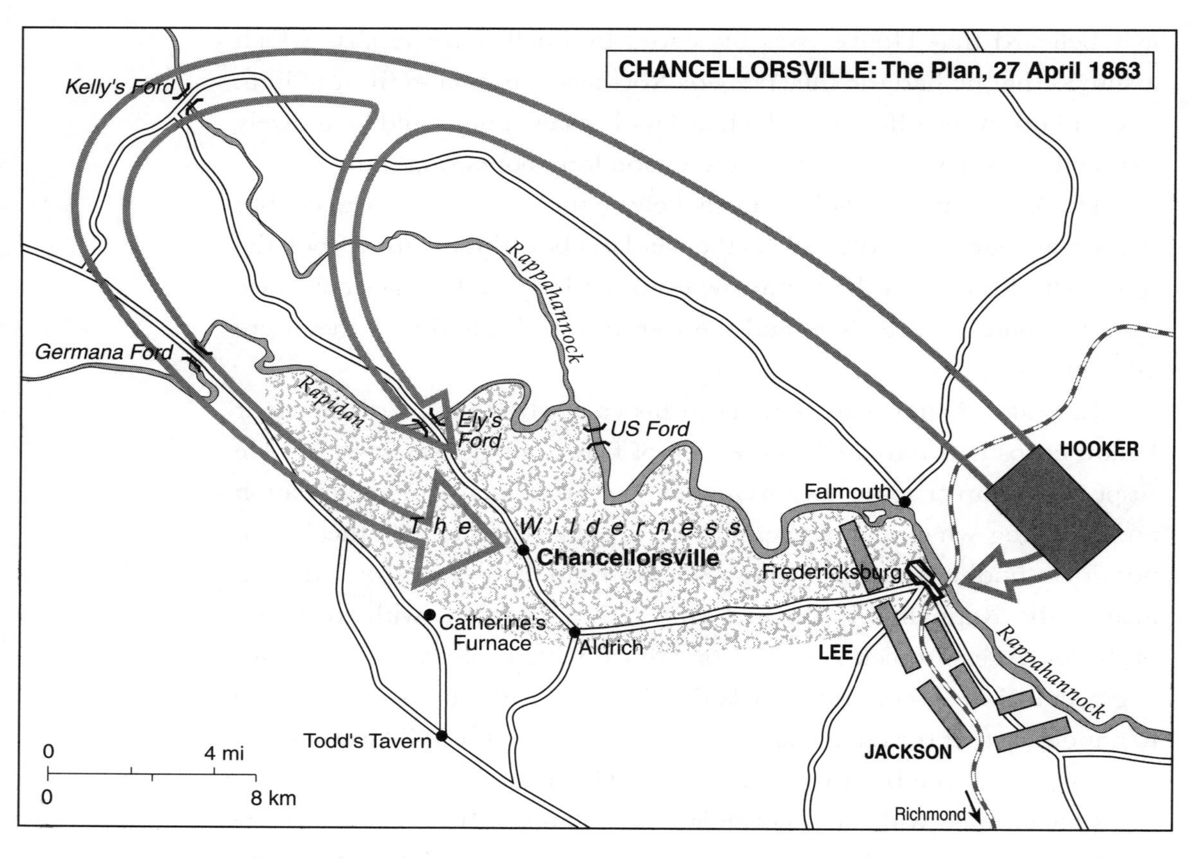

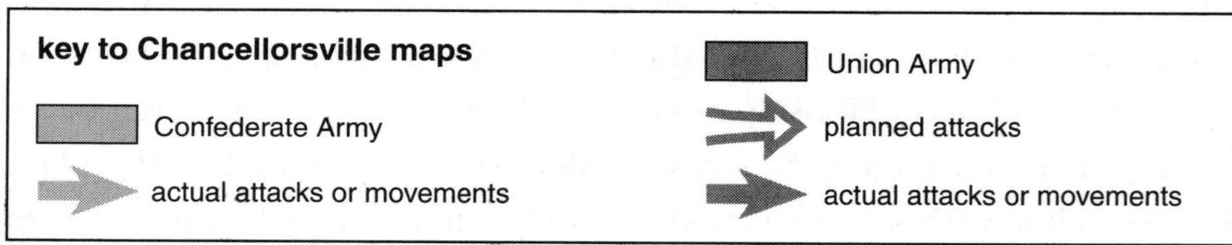

The Battle of Chancellorsville

Jackson's movements in the early morning light were seen by a Union battery which opened fire, but Jackson moved his column to another road (still known today as 'Jackson's Trail') and kept moving. In order to find out what was going on, two Union divisions from Major General Daniel E Sickles's III Corps advanced, driving back a Confederate outpost. Jackson sent back some troops which aided others already in position on the other side of the road, and the Union force was quickly ambushed and driven out. By 3.00 p.m., driving forward in the hot sun, Jackson had posi-

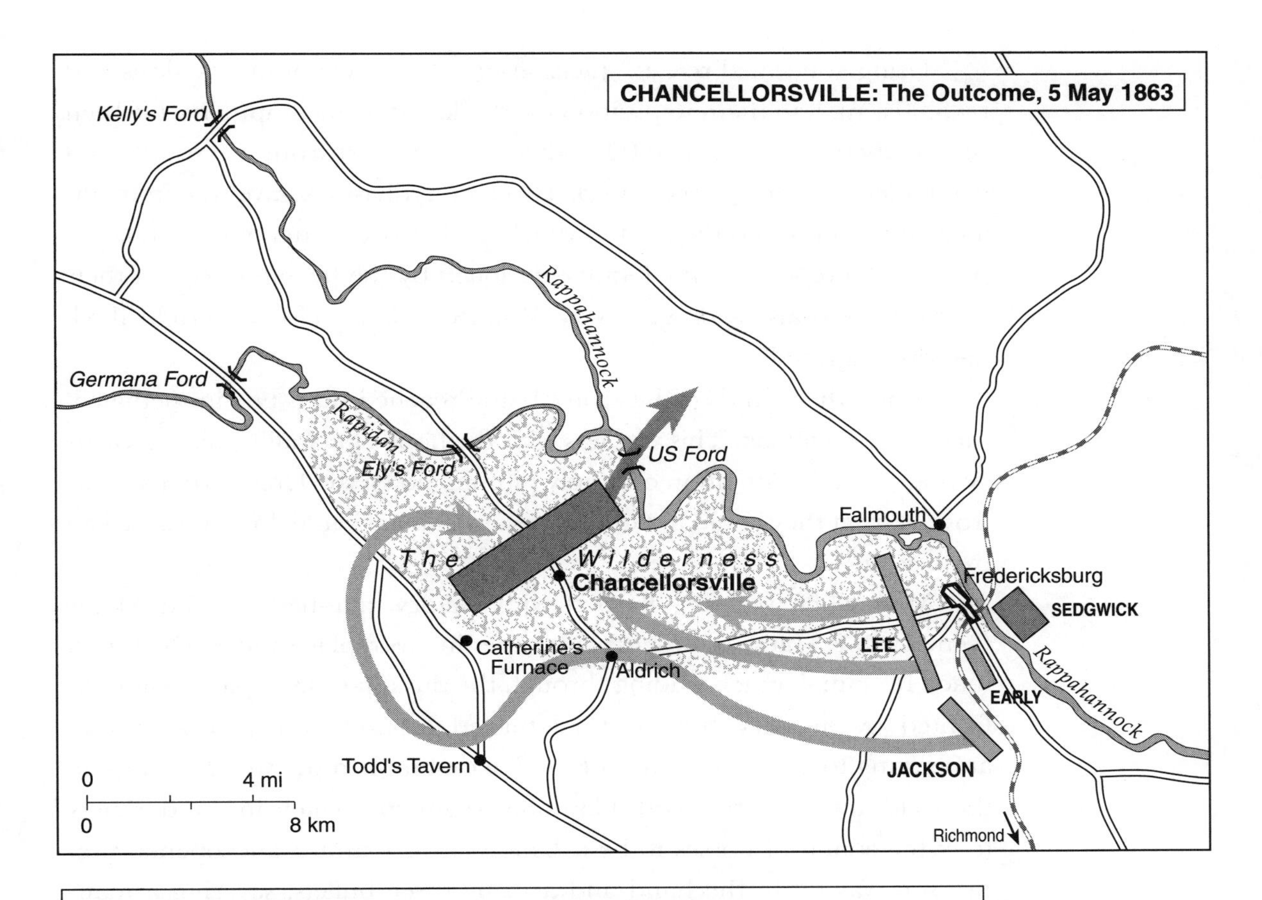

The Battle of Chancellorsville, 1–5 May 1863

The Confederate Army of Northern Virginia under General Robert E Lee
The Union Army of the Potomac under Major General Joseph E Hooker
Confederate forces: 62,000 troops
Union forces: 138,378 troops
Confederate casualties: 12,500
Union casualties: 17,300

Critical Moments

Hooker makes his flank march to Chancellorsville
Sedgwick crosses the Rappahannock east of Fredericksburg
Hooker withdraws to defensive positions in the Wilderness
Jackson makes his separate flank march onto Hooker's rear
Hooker falls back and Lee's army re-unites
Lee moves against Sedgwick
Both Hooker and Sedgwick fall back across the Rappahannock

tioned his men directly in the rear of Hooker's army. Guided by Lee's cavalryman son, Brigadier General Fitzhugh Lee, he was quietly overlooking an encampment of Union troops sprawling in the sun, smoking and eating, secure in the knowledge that they were well behind the line of battle.

Using concealed routes, Jackson moved up two of his divisions and by 5.00 p.m. had them in position overlooking the encampment. He then turned them loose, and 18,000 howling men burst from the woods and descended onto the Union camp. Two Union divisions gave way immediately and simply fled the field. A third, further back, attempted to take up positions to repel the attack but were foiled by the fleeing troops of their own side, and also soon gave way. Within an hour, Hooker's right flank had disintegrated.

When the sounds of Jackson's battle reached him, Lee immediately turned to the attack. This was less with the hope of actually gaining victory than with the intention of simply pinning down the Union troops to his front, so that they could not disengage and fall back to help those attacked by Jackson.

Meanwhile Jackson, flushed with victory, planned to send Major General A P Hill's six brigades into the line to replace those which had made his initial attack. Riding through the dusk, Jackson's party was confronted by artillery fire, so they turned around and galloped back, attempting to get out of range of the Union guns. In the woods alongside the road lay the 18th North Carolina Regiment, which in the darkness took the sound of horses to be a Union cavalry attack and opened fire. Jackson was hit in the hand and arm by three bullets, severing a main artery. He was hurried through shellfire to a field hospital where his arm was amputated.

When Lee was informed of Jackson's wounds (before the amputation was known about) tears sprang to his eyes. 'Any victory is dearly bought', he said, 'which deprives us of the services of General Jackson, even for a short time.' But in spite of his distress, he immediately turned to the problem of picking up where Jackson had been forced to leave off. A P Hill had also been wounded, and the forward troops had no inkling of what Jackson had planned. Lee therefore gave Jackson's command to Jeb Stuart, with orders to drive forward and unite the two portions of the Confederate army, since the opportunity to cut Hooker off completely had now passed. Hooker had made another error during the night, in withdrawing some of his artillery from a commanding position on a hill called Hazel Grove, just west of Chancellorsville. Stuart spotted this, and directed Confederate artillery to the position so that it could rake the Union lines. Together with spirited Confederate attacks, this led Hooker to pull back even further and to set up a defensive line so that his army now held a triangular position, with one flank resting on the Rapidan and the other on the Rappahannock. Under Stuart's inspired leadership, the Confederate forces from the west pushed forward around the apex of Hooker's triangle until they met the advance guards of Anderson's forces, which had been left behind when Jackson made his flanking movement. Lee's army was once again a single unit; his immense gamble in splitting his forces had succeeded.

The Final Moves

Hooker fell back from Chancellorsville starting on 3 May. But meanwhile the Union forces under Sedgwick at Fredericksburg had finally managed to drive the Confederates under Brigadier General Jubal Early out of their defensive positions. The Confederates fell back southwards as they had been ordered, but this allowed Sedgwick's force to move west in an effort to take the pressure off Hooker. This was successful, as Lee was forced to break off his attacks in order to send two brigades back to stave off Sedgwick's advance. Lee now determined that Hooker was too securely positioned to be moved, so he decided to finish off Sedgwick instead. Anderson was deployed around Sedgwick's flank, and Jubal Early, with the troops who had retired southwards from Fredericksburg, now returned to threaten Sedgwick's rear. Caught on three sides, Sedgwick rapidly turned about and crossed the Rappahannock to safety.

Lee now gathered his forces together and deployed them for one last drive against Hooker, but Fighting Joe did not wait to argue. During heavy rain on the night of 5 May he withdrew his troops across the Rappahannock, and when the next day dawned there was nothing in front of Lee's army except the river. Against all odds and conventional precepts of warfare, Lee had defeated a force three times his own strength. It was undoubtedly his masterpiece. Unfortunately for Lee, the pleasure was completely ruined by the death of Stonewall Jackson: after the amputation pneumonia supervened, and after lingering for a week Jackson died on 10 May. Lee had not only lost a friend, he had lost his most reliable general, a man with whom he could plan and execute manoeuvres of any degree of complexity, secure in the knowledge that his orders would be understood and carried out. Without Jackson it is doubtful if Lee could have won at Chancellorsville; and without Jackson it was almost certain that nothing so daring could ever be attempted again.

IAN HOGG

Further Reading

Earle, P., *Robert E. Lee* (London, 1973)
Parish, P.J., *The American Civil War* (London, 1975)
Selby, J., *Stonewall Jackson* (London, 1968)
Stackpole, E.J., *Chancellorsville* (Harrisburg, PA, 1958)

The Second Battle of Alamein

23 October–12 November 1942

The Proper Application of Overwhelming Force

'The commander must decide how he will fight the battle before it begins. *He must then use the military effort at his disposal to force the battle to swing the way he wishes it to go; he must make the enemy dance to his own tune from the beginning and never* vice versa. *To be able to do this, his own dispositions must be so balanced that he can utilize but need not react to the enemy's move, but can continue relentlessly with his own plan. The question of 'balance' was a definite feature of my military creed.'* Montgomery

Every commander must not only have faith in himself, but also make his army believe that it is the best in the world. In considering the classic ploys of battle, it is the spectacular manoeuvres which have always appealed to the imagination, and both commanders and their armies have been judged by their skill and training in being able to execute such manoeuvres. But what if the situation is different? Every manoeuvre carries with it a risk, and some battles are just too important to risk losing. If the enemy is known to be more skilful in manoeuvre, but also to have weaker numbers, there are times when the best battle plan is one which simply aims to crush him relentlessly with superior strength. Winston Churchill even described the Allied strategy to win World War II after 1941 as being 'the proper application of overwhelming force' in this manner. It saves lives too, on your own side. But as this famous battle shows, it is not as easy to do as it may appear.

The Background to the Battle

The 'Desert War', fought in North Africa between the Axis powers and the British (with their Allies and Empire forces), saw many changes of fortune, beginning with Lieutenant General Sir Richard O'Connor's lightning campaign of December 1940 which had almost destroyed the Italians in Libya. The German 'Afrika Korps' under Erwin Rommel (who would rise to Field Marshal in the course of the Desert War) restored the situation in the Axis' favour in the spring of 1941 with its dramatic attack through Cyrenaica. Rommel's siege of Tobruk, the westernmost supply port for the British Eighth Army, was lifted only at the end of the bloody 'Operation Crusader' battles launched by the British in November. But Rommel's retreat was only temporary, and in May 1942 he began his attack on the Gazala line defended by the British west of Tobruk. By the end of June, Tobruk had fallen, the Eighth Army was in full retreat, and Rommel was driving his men on towards Egypt and Alexandria.

Rommel's victorious forces, now known as the 'Panzerarmee Afrika', were finally halted at the First Battle of Alamein by a 'brave but baffled' Eighth Army (as Winston Churchill put it) under General Sir Claude Auchinleck during July 1942. In August, Lieutenant General Bernard Montgomery came out to command a revitalized and re-equipped Eighth Army, which easily halted Rommel's last, desperate attempt to reach the delta in the Battle of Alam Halfa, fought at the end of August. With Egypt now finally secure, and Rommel's Panzerarmee Afrika in a desperate supply situation (its main supply base and port of Tripoli was over a thousand miles away to the west), Montgomery could plan his own offensive to, as he put it, 'hit Rommel for six right out of Africa'.

Montgomery's Plan

The Alamein position was unique in the Western Desert because it had secure flanks, resting on the sea in the north and the Qattara Depression in the south. Thus the 37 miles of front could be easily held by both armies, which had concentrated their forces to hold the line. In this circumstance, the standard desert tactic of sweeping round the enemy's southern flank with mobile forces was simply inappropriate. Instead, Montgomery planned to mount a direct assault to break through Rommel's line in the north, near the coast. This would make maximum use of Montgomery's massive superiority in forces over Rommel, something which Auchinleck before him had never possessed. As well as having over a thousand guns and almost complete control of the air, the Eighth Army outnumbered the Panzerarmee Afrika by more than two to one in infantry and tanks, had generally better equipment at all levels, and was in a much better position regarding both reserves and supplies. Also, like many 'British' armies of history, Montgomery's Eighth Army was an amalgam of troops from many parts of the world, with contingents from the British Empire and various Allies.

The Axis forces on the Alamein line had dug themselves in, and planted what they called 'Devil's Gardens' of mines and booby traps up to five miles deep. To cope with this, Montgomery planned to open the battle with a massive artillery bombardment, and then to use the infantry divisions of his XXX Corps under Lieutenant General Sir Oliver Leese to make a breach in these defences. Meanwhile, sappers would 'gap' (make gaps in) the minefields for both the infantry and armour. Once the breach was made, the infantry would continue to 'crumble' Rommel's infantry formations which held the minefields, by which Montgomery meant wearing them down with attacks and firepower until they gave way. Meanwhile, the British armour of X Corps under Lieutenant General Herbert Lumsden (1st and 10th Armoured Divisions) would sally forth from the minefields to destroy the formidable armour of Rommel's army, contained in his German 15th and 21st Panzer Divisions and the Italian Ariete and Littorio Divisions. This would produce what Montgomery called a 'dogfight', a fierce battle of attrition in which victory would go to the stronger side, followed by a British 'break-out' into open desert, having blasted a hole right through the northern centre of the Axis line.

Just as importantly, given that Montgomery planned to fight a frontal battle of attrition in which the British superiority in supply and equipment would eventually tell, it was essential that Rommel was unsure of where the blow might fall. Accordingly, a sophisticated deception and misinformation plan was set in motion to persuade the Germans that the attack would be delivered in the south, with the British once more making an armoured sweep around their enemy's open desert flank, as had become almost standard for the attacking side in all previous desert battles. False radio traffic, dummy tanks and ve-hicles, and even a dummy fuel pipeline were all constructed in the southern sector. To add realism to these efforts, the troops of XIII Corps under Lieutenant General Brian Horrocks, including 7th Armoured Division (the famous 'Desert Rats') were to mount a limited offensive in the southern sector in an attempt to hold both 21st Panzer and Ariete away from the main fighting.

In its essentials then, Montgomery's plan was simple enough. He estimated that the whole process, from the first break-in, to crumbling, to dogfight, and then the break-out would take 12 days.

Rommel and Stumme's Plan

Rommel was an ill and disheartened man by September 1942. He flew home to Germany for treatment early that month and left his Panzerarmee in the hands of General Georg Stumme, an experienced armoured commander. However, Rommel had worked out his defensive plan for the battle he knew was coming. With general supply shortages and a crippling lack of petrol, there was little Rommel could do. His army could neither advance nor retreat for lack of fuel, which meant that it had to wait and meet the British attack. Nonetheless, Rommel had disposed his infantry formations in depth, corseting the weaker Italian formations in between German ones, with outposts within the minefields themselves. His armoured formations were held back, with 15th Panzer and Littorio in the north, 21st Panzer and Ariete in the south, and the German 90th Light Division in reserve. Rommel's basic plan was to absorb the British blow, making any advance as difficult as possible. However, his real difficulty lay in the fact that there was only enough petrol to bring 21st Panzer and Ariete up from the south once. There was not enough fuel to send them back, so Rommel had to be certain where Montgomery's main attack was being made before he committed his armour. Once he had identified the British *schwerpunkt* (a German term meaning 'the critical point of an attack') Rommel would concentrate his armour and counter-attack with all his might to seal off any breach. With this accomplished, Rommel could only hope that the front would bog down into stalemate once again.

Operation Lightfoot

The Eighth Army attack, code-named Operation Lightfoot, began at 9.40 p.m. on 23 October with the biggest artillery bombardment since World War I. Nearly all Montgomery's thousand guns opened up along the front, causing great disruption and confusion to the Axis defenders. The opening of the offensive had caught the Panzerarmee Afrika by surprise and the initial Axis response was weak and slow. Initially, the soldiers of the 9th Australian Division, 51st (Highland) Division, 2nd New Zealand Division, and 1st South African Division, each attacking on a two brigade front, made good progress through the enemy positions and minefields. However, once the Germans and Italians began to recover from the shock, the fighting became bitter and confused all along the front. Advancing in darkness and with huge clouds of dust obscuring their vision, it became increasingly difficult for the attacking troops to keep direction, and mines and booby traps caused considerable casualties. While the Australians and New Zealanders were able to capture their final objectives, the Highlanders were stopped short of their object-ive, known as the Oxalic line, except on the extreme left flank, while the South Africans were halted quickly and suffered heavy casualties. Nonetheless, the infantry attack had bitten deeply into the Axis lines and caused considerable disruption.

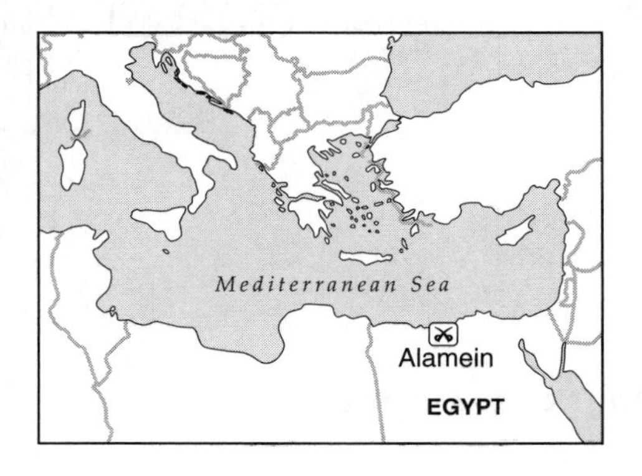

While the infantry attack was proceeding, the sappers and pio-

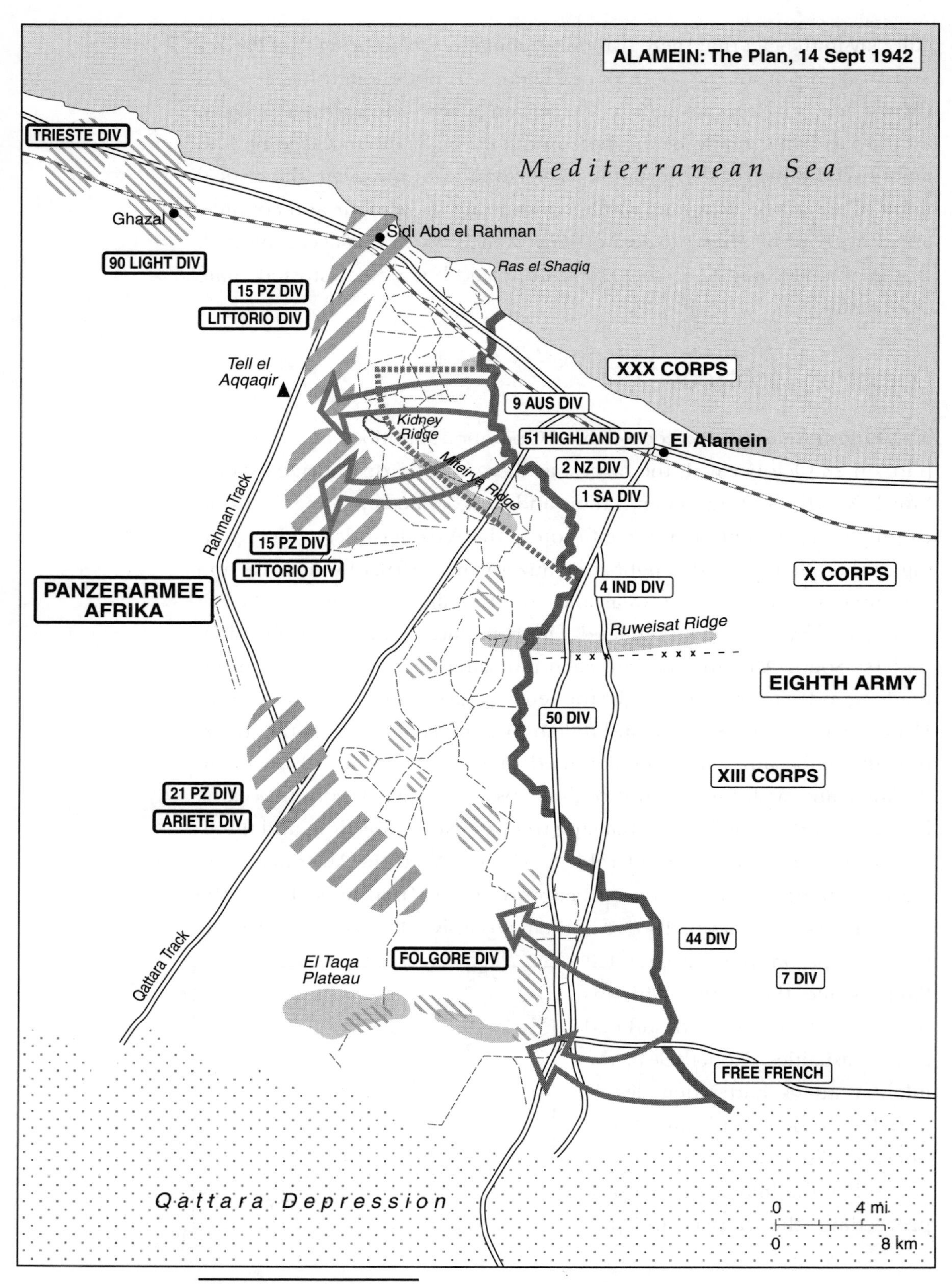
ALAMEIN: The Plan, 14 Sept 1942
Mediterranean Sea
TRIESTE DIV
Ghazal
Sidi Abd el Rahman
Ras el Shaqiq
90 LIGHT DIV
15 PZ DIV
LITTORIO DIV
Tell el Aqqaqir
XXX CORPS
9 AUS DIV
Kidney Ridge
51 HIGHLAND DIV
El Alamein
Miteirya Ridge
2 NZ DIV
1 SA DIV
Rahman Track
15 PZ DIV
LITTORIO DIV
PANZERARMEE AFRIKA
4 IND DIV
X CORPS
Ruweisat Ridge
EIGHTH ARMY
50 DIV
XIII CORPS
21 PZ DIV
ARIETE DIV
44 DIV
Qattara Track
El Taqa Plateau
FOLGORE DIV
7 DIV
FREE FRENCH
Qattara Depression
0 4 mi
0 8 km

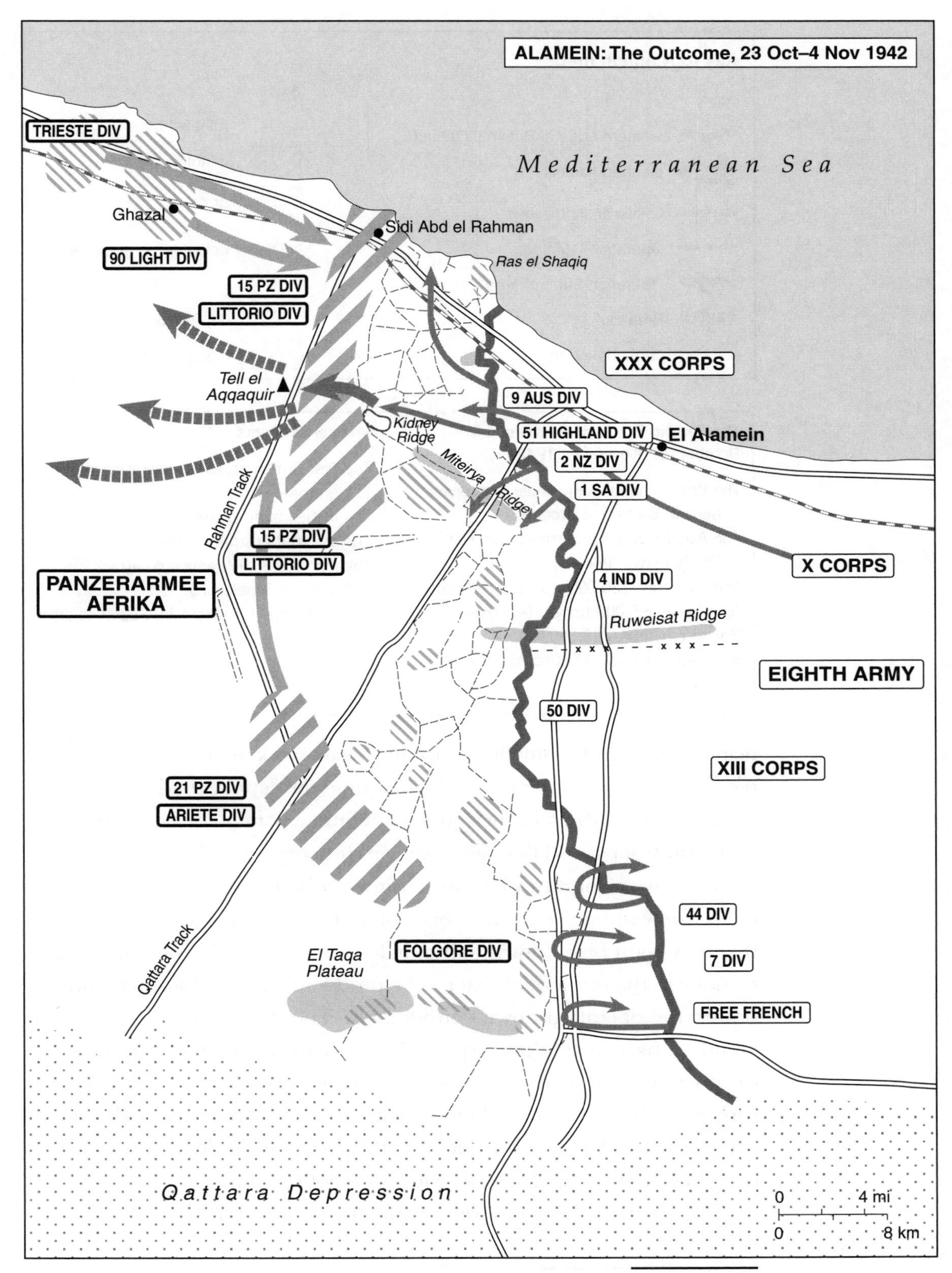

ALAMEIN: The Outcome, 23 Oct–4 Nov 1942
Mediterranean Sea
TRIESTE DIV
Ghazal
Sidi Abd el Rahman
90 LIGHT DIV
Ras el Shaqiq
15 PZ DIV
LITTORIO DIV
XXX CORPS
Tell el Aqqaquir
9 AUS DIV
Kidney Ridge
51 HIGHLAND DIV
El Alamein
Miteirya Ridge
2 NZ DIV
1 SA DIV
Rahman Track
15 PZ DIV
LITTORIO DIV
X CORPS
PANZERARMEE AFRIKA
4 IND DIV
Ruweisat Ridge
EIGHTH ARMY
50 DIV
XIII CORPS
21 PZ DIV
ARIETE DIV
44 DIV
Qattara Track
El Taqa Plateau
FOLGORE DIV
7 DIV
FREE FRENCH
Qattara Depression
0
4 mi
0
8 km

key to Alamein maps

Allies		Axis	
(symbol)	western edge of British minefield	(symbol)	minefield
(symbol)	planned attacks	(symbol)	infantry
(symbol)	Oxalic final objective	(symbol)	armour
(symbol)	Operation Lightfoot	(symbol)	actual troop movements
(symbol)	Operation Supercharge		
(symbol)	Breakout	(symbol)	depression
- x x x -	corps boundary	(symbol)	high ground

The Second Battle of Alamein, 23 October–12 November 1942

The British Eighth Army under Lieutenant General Bernard Montgomery
The Axis Panzerarmee Afrika under Field Marshal Erwin Rommel
British forces: 220,000 troops, 1,100 tanks
Axis forces: 96,000 troops, 500 tanks
British casualties: 13,560
Axis casualties: 61,000 (including 35,000 prisoners)

Critical Moments

The first British attack in Operation Lightfoot
The British feint in the south
The renewed Australian attack northwards
The renewed British armoured attack in Operation Supercharge
The final British breakout into open country

neers tasked with minefield gapping were working hard to make the required number of gaps. For the first time the sappers had a number of gadgets to help them in their job, including new mine detectors supplied by Polish troops, and flail tanks known as Scorpions. However, the mine detectors were found to be too sensitive to be useful, and the Scorpion tanks threw up such choking clouds of dust – as well as exploding mines – that they soon broke down. Ultimately, most of the mine clearing had to be done in the traditional manner by prodding with the bayonet, and this slowed the operation down considerably. The Axis minefields were also found to be deeper than expected, which meant that by dawn on 24 October there were fewer gaps through the minefields than had been expected. On 1st Armoured Division front only one gap was made, but on 10th Armoured Division front with the New Zealanders four gaps had been made, up to the Miteirya Ridge but not beyond.

The men of the armoured regiments had spent a distinctly unpleasant night on 23 October, moving up nose to tail along tracks which were feet deep in dust. Delays caused by dust, traffic jams, and misdirections meant that 10th Armoured Division did not get forward as quickly as had been hoped, although units of the 9th Armoured Brigade, which had been attached to 2nd New Zealand Division did manage to reach the Miteirya Ridge. At dawn on 24 October, the elements of 10th Armoured Division which had managed to get through the minefield gaps attempted to advance off the Miteirya Ridge and found themselves exposed to the full weight of the Axis antitank gunners and artillery. Meanwhile, many other tanks were stuck in traffic jams in the minefields themselves or further back. It was an impossible situation for armour. Major General A H Gatehouse, the 10th Armoured Division command-er, expressed these views to Montgomery in a heated exchange over the telephone early on 25 October but Montgomery flatly ordered him to push on with his armour regardless of casualties. Nonetheless, after a day of battle where the exhausted infantry had to crouch in their hastily dug foxholes as an armoured battle raged around them, Gatehouse ordered his tanks to withdraw back to the ridgeline. Montgomery's planned break-out had failed. After the event, it was clear that the minefields were too deep to be penetrated in one night attack, and that the armour would be left penned in to face the teeth of the Axis defence.

On 25 October, feint attacks were mounted by 7th Armoured Division, supported by 50th (Northumbrian) Division on the southern sectors of the front. However, rough terrain, extensive minefields, and the spirited defence by the Italian Folgore Parachute Brigade meant that 7th Armoured Division suffered heavy casualties for little gain. Montgomery wished to keep this division as what he described as 'a force in being' and the attack was called off on 27 October. Nonetheless, these efforts had done their job: unsure of the main thrust, 21st Panzer and Ariete were still held in the south, away from the main battle.

Axis uncertainty was compounded by confusion in their chain of command. When the battle began, General Stumme was in command of the Panzerarmee. His most controversial decision was to forbid retaliatory artillery fire when the British barrage opened. However, on 24 October, he decided to go to the front lines to see what was happening. Unfortunately, his car was caught in an artillery barrage and Stumme, who was about to climb out, had a heart attack and was flung from the car as the driver sped off. Stumme's fate was not learnt until the next day, and in the meantime General Wilhelm Ritter von Thoma took command, and Rommel was recalled from Germany. Rommel arrived with his forces at Alamein late on 25 October, but it took time for him to understand the confused situation, thus delaying the Axis reaction still further.

The Dogfight

Montgomery's initial break-out plan had not succeeded, and on 25 October he cancelled any further attacks by 2nd New Zealand Division, and pulled out 10th Armoured Division to rest and refit near Alamein station. Instead, Montgomery ordered an attack northwards by 9th Australian Division, supported by 1st Armoured Division. This attack began that night, and met with initial success, threatening to cut off elements of the German 164th Light Division around Ras el Shaqiq. While this northern attack got underway, Montgomery continued his ruthless application of superior numbers and weight of firepower at other points of the front. Heavy artillery bombardments, air attacks and limited infantry attacks were all thrown at the Axis defenders during the next few days in an attempt to wear down their strength. These crumbling attacks, while expensive for the British, were successful in writing down Rommel's strength and pinning his main forces. Perhaps the best example of these operations was the action around the location code-named 'Snipe', fought near the centre of the battlefield on 27 October by the 2nd Battalion of the Rifle Brigade (commanded by Colonel 'Vic' Turner) and 239th Antitank Battery of the Royal Artillery. During the night, this small force had advanced right into the Axis positions. Next day, the men of the Rifle Brigade and Royal Artillery struggled to man their nineteen 6-pounder antitank guns in the face of numerous German and Italian tank attacks, as Rommel pushed his armoured divisions in a counterattack aimed at sealing the breach in his line. By evening on the 27 October, Turner and his battalion were forced to withdraw owing to lack of ammunition, but during the day his men had managed to knock out over 34 Axis tanks and armoured vehicles, a significant proportion of Rommel's remaining tank strength.

On the night of 30–31October the Australians launched another attack towards the coast. Although they were met with heavy fire, particularly from a strong point known as 'Thompson's Post' and a blockhouse near the railway line, the Australians managed to get across the railway and coast road, thus endangering the 164th Light Division. Rommel's attention was drawn to the sector and he personally supervised an attack by 90th Light Division and elements of 21st Panzer Division on 31 October in an attempt to stabilize the situation. Constantly hammered by Axis artillery and forced to repel numerous German attacks, the Australians suffered enormous casualties but were able to hang on in the face of daunting odds and bitter attacks.

Operation Supercharge

While the Australians were drawing everything onto themselves, Montgomery again changed the direction of his main thrust by planning Operation Supercharge. This was to be the final break-out, using a modified version of Operation Lightfoot. An initial infantry attack would be supported by an enormous concentration of guns, which would then make a breach around Tell el Aqqaqir, about 15 miles west of Alamein station on the Rahman track, for the armour to pass through and complete the destruction of the Panzerarmee Afrika. Montgomery entrusted the task to Major General Bernard Freyberg, commander of the 2nd New Zealand Division, although the infantry for the attack actually came from two British brigades. Postponed for 24 hours owing to the complexity of the planning and practical difficulties, Operation Supercharge opened on the early hours of 2 November 1942. At 1.05 a.m. shells from over 800 guns crashed out on a frontage of only 4,000 yards. With this huge weight of artillery support, the attack of the two infantry brigades went like a drill, and they achieved their objectives deep within the Axis positions without too much difficulty. Montgomery had informed Brigadier John Currie, commander of 9th Armoured Brigade, that he was prepared to accept 100% losses in his formation in order to achieve success. Currie's job was far from enviable. His brigade was to assault the German artillery and antitank gun screen just before dawn and thus create a breach, which the regiments of 1st Armoured Division could then exploit. Unfortunately, Currie's attack was delayed by half an hour, so that although his tanks did reach the German gun line and manage to destroy over 30 guns, his tanks were caught silhouetted by first light, and thus became perfect targets for the German 88 mm antitank gunners. Literally in minutes Currie's brigade was eliminated, and by the end of the day he had lost 75 tanks out of 94.

1st Armoured Division was also delayed in its approach march, which meant that it was unable to exploit the confusion caused by the sacrifice of 9th Armoured Brigade. But using the gap in the German gun screen, its regiments took up defensive positions under the protection of the British artillery. Rommel's instinctive reaction to counterattack the penetration followed, and 1st Armoured Division was attacked by both the 15th Panzer Division and 21st Panzer Division, and 90th Light Division, as well as elements of the Italian Littorio Division and motorized Trento Division. Desperate Axis counterattacks went on for the rest of the day, and the confusion and carnage of this tank battle was enormous. 'For hours', one survivor described it, 'the whack of AP [armour piercing] shot on armour plate was unceasing.' The British armour stood its ground and inflicted heavy losses on the Germans; 117 Axis tanks were destroyed during the battle. The grim reality of the situation dawned on Rommel that evening, when von Thoma reported that his German forces had been reduced to one third of their initial strength and had only 35 'runners' (functioning tanks) for 3 November. The British break-out had been held but at enormous cost. However, that night Rommel ordered a withdrawal along the coast to Fuka, 60 miles to the west of Alamein.

The End of the Battle

3 November was a frustrating day for the British, as they tried without success to push through the battered Panzerarmee, seeing that Rommel's troops were visibly thinning out behind a powerful rearguard and gun line. But as this withdrawal got under way, Rommel received a shocking order from Hitler, who ordered him to stand fast and insisted that not a yard of ground was to be given up. Astounded but obedient, Rommel countermanded as many of the retreat orders as possible. This caused enormous confusion amongst the Axis units, and particularly condemned the Italian infantry units in the south to destruction and capture.

On 4 November, British armoured cars of the 9th Royal Lancers finally squeezed past Axis units and broke out into the open desert, quickly followed by other units of the Eighth Army. Field Marshal Albert Kesselring, the German Commander in Chief South, personally countermanded Hitler's orders and allowed Rommel to retreat once more. But as Rommel desperately attempted to save as much of his army as possible, the Eighth Army was able to bag large numbers of Axis prisoners in a series of sharp actions. The battle of Alamein was over.

Montgomery's plan had not gone as predicted, but the battle had lasted for 12 days. The Eighth Army had suffered 13,560 casualties and lost 500 tanks (300 of which were recoverable) but Rommel's Panzerarmee Afrika had been smashed and over 35,000 Axis prisoners were captured. When the Anglo-American landings in Morocco and Algeria, code-named Operation Torch, took place on 8 November, the Axis position in North Africa was doomed. While Montgomery has been criticized for his conduct in the pursuit of Rommel's fleeing troops, by 23 January 1943 Tripoli, the last city of the Italian Empire, had fallen, and the race into Tunisia continued. The Desert War was over, and in Churchill's words, 'It might almost be said that before Alamein we never had a victory, while after Alamein we never had a defeat.'

NIALL BARR

Further Reading

Barnett, C., *The Desert Generals* (London, 1960)
Carver, M., *El Alamein* (London, 1962)
Macksey, K., *Rommel, Battles and Campaigns* (London, 1997)
Hamilton, N., *Monty, The Making of a General* (London, 1981)

Surprise

'Whatever a thing may be... the less it has been foreseen the more it pleases or frightens. This is seen nowhere better than in war, where surprise strikes with terror even those who are much the stronger party.' Xenophon

A plan that can surprise the enemy is one that will probably defeat him in battle. But in war no one knows what the enemy will really do, and surprise can take many forms. Attacking at a place or a time that the enemy does not expect will always give the attacker the advantage, but there are also ways to surprise the attacker with an unusual form of defence, or to achieve surprise by methods that have little to do with showing up somewhere unexpectedly. It is always easy, after a successful battle that has depended for its success on surprise, to see how the surprise has worked. At the time it is not so obvious!

Crécy

26 August 1346

Surprise Against the Odds

'The English archers then advanced one step forward, and shot their arrows with such force and quickness, that it seemed as if it snowed... ' Froissart

If it is hard to be certain after the event if any battle really went according to plan, whatever might be claimed by the victorious leader, then it is almost impossible for the battles of the Middle Ages. The medieval chroniclers, with their very different traditions of recording events from our own, seldom allow us to be certain that what we are hearing are the real voices of the commanders as they plan their battles. But it is certainly true that some medieval armies, at least, were far from being clumsy hordes of men, and that some commanders had too many victor-ies against the odds for sheer luck to be responsible. The stereotype of medieval generals thinking only of cavalry charges is not true, and chivalric commanders did not always opt for all-out attack. This famous battle from the Hundred Years' War shows that medieval kings could also be gifted generals, and knew how to spring a surprise or two.

The Military Background

What happened in 1346 resulted from circumstances specific to the contestants and the campaign. When Edward III of England invaded France in 1346, he was claiming the French throne. He was also attempting to bring the French king to battle. He had twice failed to draw Philip VI of

France into a fight in Flanders half-a-dozen years earl-ier, and the cost of those expeditions, involving as they did an ambitious political alliance against France, had beggared the English government. During the 1339 campaign, Edward had offered battle at La Capelle. But Philip decided not to attack a strong defensive position, and instead withdrew into an entrenched camp at Buironfosse, a couple of miles away. Then it was Edward's turn to avoid battle.

Following his failure to coordinate attacks against the French kingdom from the northeast, Edward turned his attention to the succession dispute in the Duchy of Brittany, choosing to support John de Montfort against Charles of Blois, the French king's claimant. The English force under the Earl of Northampton, dispatched in July 1342, was quite small, with only 1,350 men divided equally between men-at-arms (a term which covers all troops who wore armour and usually rode horses but might fight either mounted or dismounted, including knights and noblemen) and archers (who might also ride to battle, but almost never fought mounted). Reinforcements under Robert, Count of Artois added 800 more troops, together with an unknown number of Bretons. Opposing them, Charles of Blois led 3,000 men-at-arms, 1,500 Genoese crossbowmen, and several thousand French levy foot (impressed peasant infantry without armour and usually of poor fighting quality).

The drawn engagement at Morlaix on 29 September which ensued was a lesson for the French, had they chosen to learn from it. Northampton dismounted his men behind a stream with a wood at their back, and further strengthened the position with ditches and pits concealed under branches. Confident in their superior numbers, the French advanced in three lines, the first two composed of mounted men-at-arms, the rear one of men on foot. Led by Geoffrey de Charny and his newly formed chivalric Order of the Star, the French first line charged and was thrown into confusion by a combination of the field defences and English archery. The second line came on too soon to see what was happening, and joined in the confused mêlée which ensued. Casualties were heavy for such a brief encounter: 50 French knights were killed and 150 captured, including de Charny. Essentially, the battle was over, but the large numbers of unengaged French footsoldiers still posed a threat to Northampton's small force. So he withdrew into the woods behind his position, and was effectively trapped there for two days, until slipping away by night. The lesson for the future was clear: the English, relying on a strong position, field defences, and their archers, could not be assailed frontally without disaster resulting. Yet, if infantry and cavalry were used in conjunction then the English tactical system could be defeated.

The Crécy Campaign

King Edward landed in western or Lower Normandy on 11 July 1348 with perhaps between 10,000 and 15,000 men. He then conducted the type of campaign which contemporaries called a chevauchée, that is to say a kind of large-scale raid, with the intention of damaging as much of the French king's territory as possible. This then undermined Philip's subjects' faith in his ability to protect them, and laid the groundwork for eventual conquest. Just how destructive this activity could be is shown by Edward's chevauchée in the Cambrèsis (the area around the town of Cambrai near the River Somme) in 1339. In a medieval version of humanitarian aid, a papal envoy sent on a charitable mission to the area in the following year recorded almost 150 destroyed towns and villages in an area only 30 miles long by 12 miles wide.

Generally, it was not the policy to engage in sieges whilst on chevauchée, since this slowed down speed of movement and caused supply problems for forces which otherwise lived off the country. But Edward's men did storm the important town of Caen in Lower Normandy. Attacking it from the land and from boats on the river, they sacked the town and massacred between 2,500 and 5,000 citizens, whilst being unable to make any impression on the powerful castle above it. This must also have been a salutary lesson to the French. Edward then advanced toward Paris, although his strategic intentions at this stage remain far from clear, and historians are divided as to what his plan actually was. He paused at Poissy some 12 miles to the north, where his men drove off a French covering force on the right bank of the Seine and constructed bridges to enable them to cross. Philip had meanwhile fallen back on Paris to gather his forces. Perhaps he was happy to see the English march northwards on 15 August. Edward may have been intending to link up with his Flemish allies, but they had been detained by besieging Béthune, 125 miles away. Philip soon set off in pursuit, as was his duty, to clear his realm of invaders.

Edward's Strategy

Edward was now in a difficult position, for as he reached the River Somme, he could not cross it in the face of determined opposition. Possibly the English chevauchée strategy was always intended to bring about a battle, but if so then Edward had put himself at a strategic disadvantage, potentially pinned against a river line and outnumbered in men-at-arms. Certainly the move north was made by forced marches on both sides. Probably at this stage, Edward had no clear, pre-determined strategy and was improvising as he went along. Certainly the medieval chronicles make a great deal of the crossing of the Somme, stressing how crucial it was to the outcome of the campaign.

Edward reached Airaines on 21 August, and sent out scouts to test the French defences. The towns of Abbeville and Amiens had bridges across the Somme, but were also fortified. Attempts to force a passage the following day were all driven back. Edward's last chance was to try the ford at Blanchetacque near the mouth of the Somme. Nowadays, the river is canalized, but then it was a shallow marsh some two miles wide, and Blanchetacque was so called because there was a white patch of chalk on the north bank which served as an aiming point for anyone crossing, in order that they did not stray from the solid causeway. There is further evidence of Edward's lack of planning, for he was apparently unaware of this ford until a local man informed him of it.

Philip seemed to have Edward in a trap. He had reached Abbeville by 23 August with the main army. The ford was defended by a reliable commander, Godemar du Fay, with 500 men-at-arms and 3,000 infantry including crossbowmen. But Edward stole a march by slipping away from Oisemont, just after midnight on 24 August. Having arrived at the ford, he had to wait for the tide to go out in full view of the French defenders, owing to the early sunrise. At about 8.00 a.m.two veteran English commanders, the Earl of Northampton and Reginald de Cobham, led 100 men-at-arms and 100 archers to force a bridgehead. Remarkably, the archers' shooting drove back the French enough to allow the men-at-arms to gain a foothold, which was then exploited by the main body. The whole English force got over in 90 minutes. They needed to, for Philip's army arrived soon afterwards; but the tide was already rising, and Philip could not cross in pursuit.

The English Deployment

This feat of arms in crossing the Somme enabled Edward to withdraw to a strong hilltop position at Crécy-en-Ponthieu, some five miles to the north. This was in territory well known to him – family lands inherited through his grandmother. It is even possible that the site had been carefully prepared beforehand, as regards supplies and ammunition provided through the port of Le Crotoy on the eastern side of the mouth of the Somme. Edward had perhaps 2,000 to 3,000 men-at-arms, 5,000 to 10,000 archers (probably towards the lower end of the scale), and 1,000 or so Welsh and Irish spearmen, and he was accompanied by many noblemen including his eldest son the Prince of Wales.

With a wood to his rear, Edward further strengthened his position with some sort of defences. The contemporary chronicler Geoffrey Le Baker says that the field was sown with pits or 'pottes', writing that, 'The English ... quickly dug many holes in the earth in front of the first line, one foot deep and one foot wide, so that if it happened that the French cavalry were able to attack them, the horses might stagger because of the holes'. Unfortunately, no trace of these pits remains at Crécy, and much about our understanding of the battle depends on where they were, and where we believe the archers to have been placed. There had been no rain for six weeks and the ground must have been pretty hard, even if it was regularly cultivated. There must also be some doubt as to whether the English army was equipped with proper digging tools. This, combined with the shortage of time available for preparation, must have resulted in pretty sketchy defences. But in addition to the wood behind Edward's position, he used baggage wagons and the horse lines to cover the army's flanks and rear, which formed a barricade to prevent encirclement by a mobile, mounted enemy. This defence had been used successfully against French cavalry a generation earlier in 1304, by Flemish forces at Mons-en-Pévèle. The size of such an encampment should not be underestimated. Even a lightly equipped force such as Edward's, which may have disposed of much of its baggage, would have had thousands of horses and scores, if not hundreds, of wagons and carts. But as regards the deployment of men, the English line could not have stretched the mile from Crécy to Wadicourt, since this would also have stretched Edward's resources of manpower too thin, and instead some wagons may have covered this exposed left flank.

Some of these carts seem to have mounted simple cannon. Certainly, the Italian chronicler Villani mentions them. If they were used, it is possible that they delivered only one discharge, more for the shock value of the sound than as killing machines.

The current historical orthodoxy on how Edward's archers were deployed is that they stood on the flanks of each 'battle' (formation) of men-at-arms, sloping slightly forward in order to provide a crossfire in front of the main battle line. Some military historians have elaborated this into a formation with projecting teeth of hollow wedges where two 'battles' joined. There is a problem with this idea, as it actually produces weak points in the English line, where, if contacted by heavily equipped men-at-arms, the archers would have been hard-pressed to defend themselves. In answer to this criticism, proponents of the idea suggest that the impact of the English archery would be to drive off attackers and funnel them into positions opposite their own men-at-arms, against whom the opposing men-at-arms preferred to fight, for reasons of social status. This argument held the field until recently, when it was pointed out that the sources actually place archers on the flanks of the army in all descriptions. The chronicler Geoffrey Le Baker is most explicit about the deployment, writing that 'The archers were arranged in their places on the sides of the king's army like wings, not in front of their men-at-arms so that they would not get in the way nor face the attack of the enemy, but shoot their arrows from the flanks'. In fact, it seems that on most occasions the English took care to protect their front with ditches or potholes, suggesting that they did not trust to hold off an enemy by firepower alone. The archers were not necessarily tied to these positions, either. They can be understood as operating like the light infantry of later periods, even skipping between the potholes, while cavalry were brought down by them. At Crécy, the archers seem to have been deployed forward and on the flanks. Their crossfire may have only covered the front of their own battle, although some in the centre may have been able to shoot over the heads of their men-at-arms owing to the terraced nature of the hillside.

The French Plan of Attack

Whatever was the case with the English deployment, the French attack failed through lack of coordination. King Philip VI is rarely given any credit for generalship, but it is worth pointing out that he had successfully defeated a Flemish force at Cassel in 1328 with a well-judged cavalry flank attack. Further, his avoidance of battle at Buironfosse in 1339, and the following year at Bouvines, had proved masterstrokes in that Edward's campaigns collapsed as a result. Such a policy took some nerve to carry through, as it meant accepting the ravaging of his lands without reply, and enduring the taunts of chivalrous young nobles that this was the behaviour of the fox and not the lion.

When Philip came upon Edward's army on 26 August 1346, he may have thought that he finally had the English at a disadvantage. He may not have reckoned that the impact of the English archers would be so significant, and in any case he had a strong force of missilemen of his own. Although it is hard to be certain of the size of any medieval army, Philip probably outnumbered Edward by at least two to one, and by perhaps a six to one superiority in men-at-arms. His forces may have numbered 12,000 men-at-arms, with some 6,000 crossbowmen, and more infantry making up some 20,000 to 25,000 men. His dispositions – if he had any in the modern sense – involved deploying the Genoese crossbowmen in front, while mounted men-at-arms formed the traditional three 'battles' in the centre, with any infantry on the flanks. But this may be all too neat a description of a force hastily deploying from line of march. Certainly the Genoese crossbowmen suffered from the lack of their 'pavises' (tall shields which protected them whilst they reloaded), as these were on carts in the baggage train. The crossbowmen have been much reviled and hence misunderstood. The chronicler Froissart suggests that they formed up under command and advanced with 'three great shouts' to keep them in formation. That they were outshot was a function of their smaller numbers and more rapid shooting of the archers, both of which might have been remedied by the pavises. A shower of rain before the battle, which dampened and slackened the cords of their bows did not help. In contrast, the English archers simply slipped off their bowstrings and hid them under their hats!

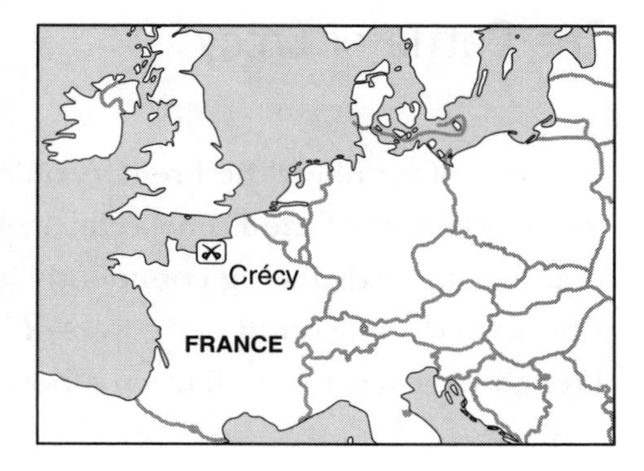

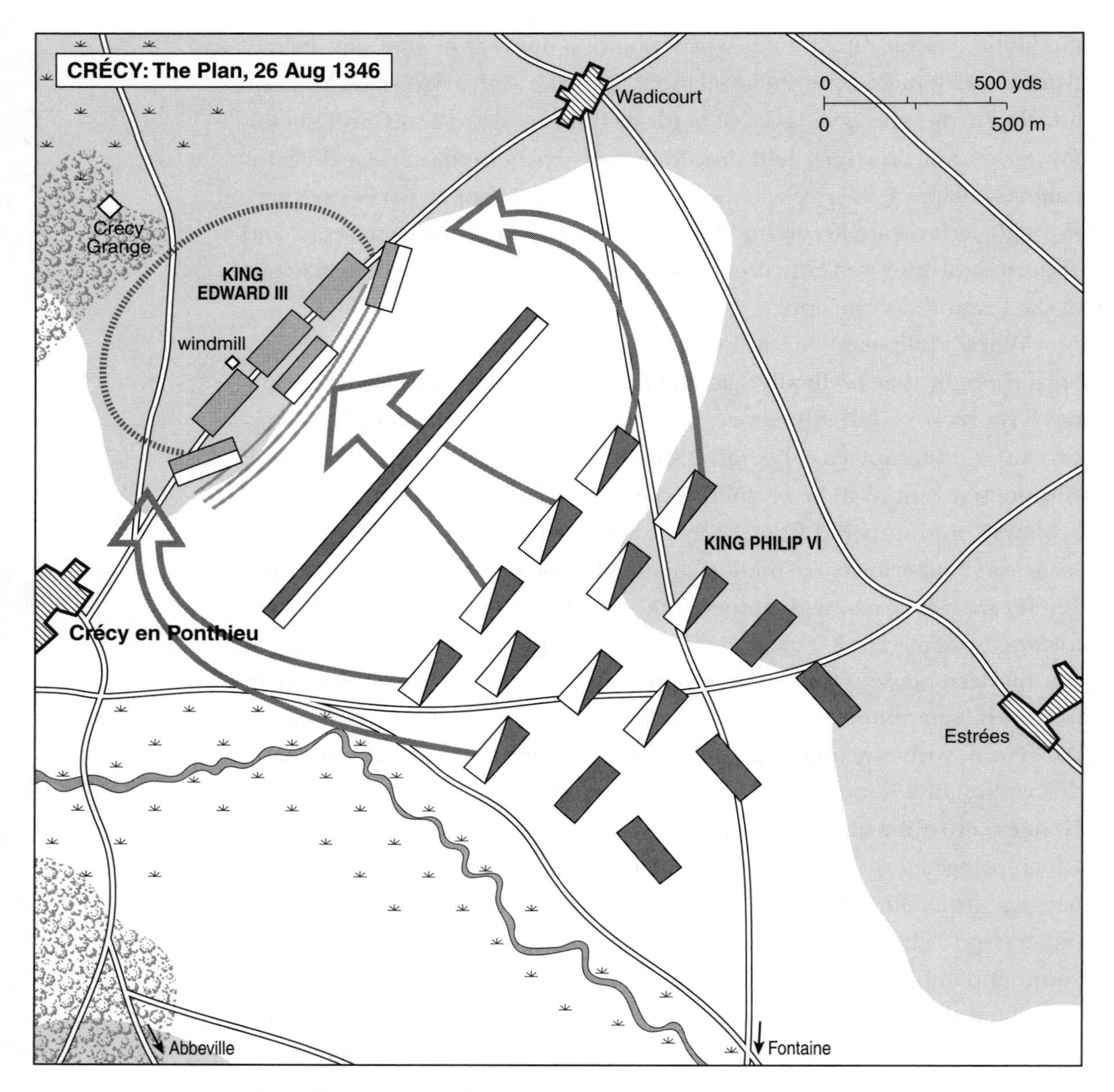

The Battle of Crécy

In terms of numbers the French could have expected a victory. But it was the impatience of the French chivalry to be at the English which was the real disaster. Advancing confidently, they ran into the double surprise of Edward's defences and the power of his archery. Froissart writes that the French believed the Italian crossbowmen to be cowards or traitors, and

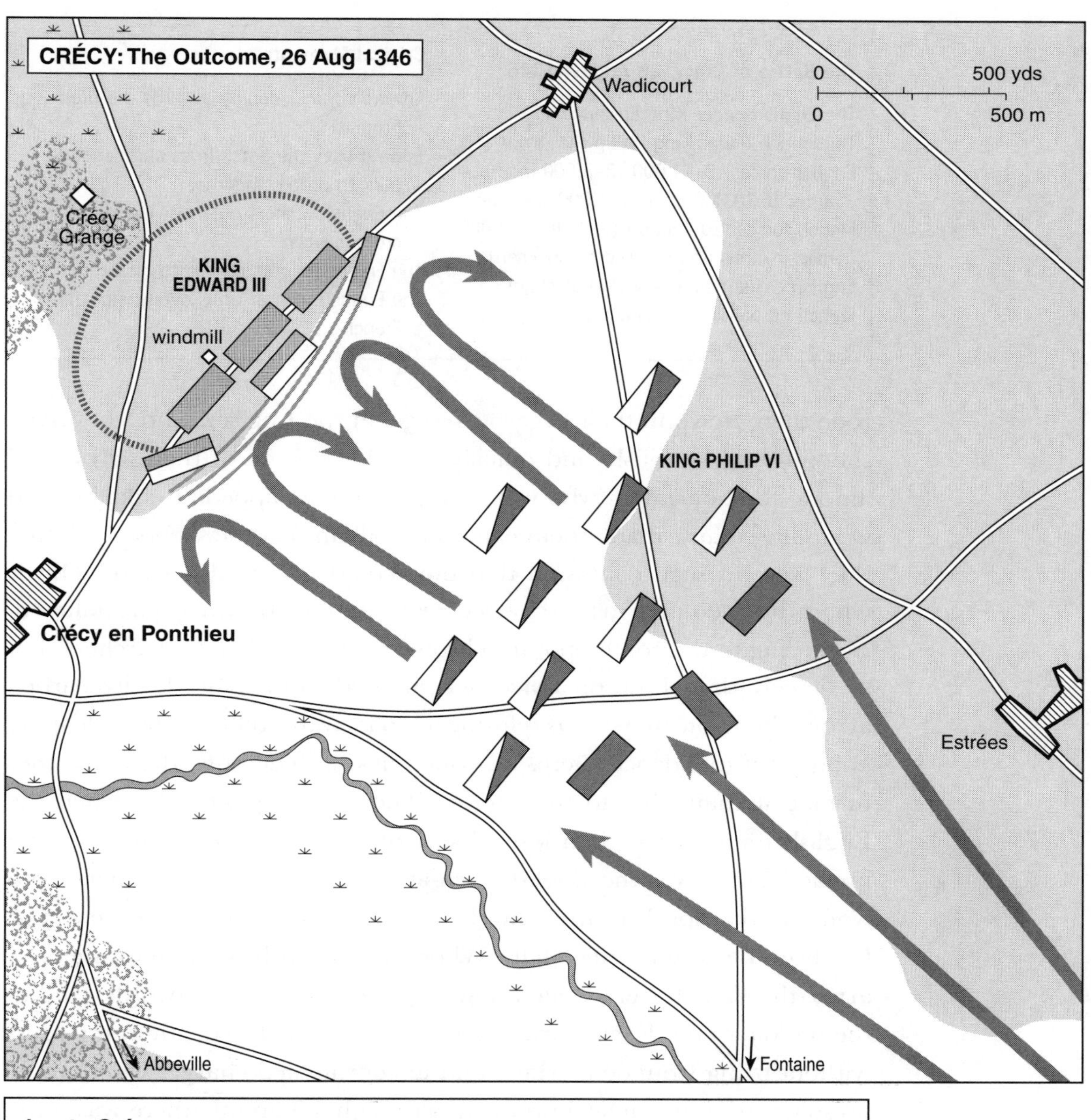
CRÉCY: The Outcome, 26 Aug 1346
Wadicourt
0
500 yds
0
500 m
Crécy Grange
KING EDWARD III
windmill
KING PHILIP VI
Crécy en Ponthieu
Estrées
Abbeville
Fontaine

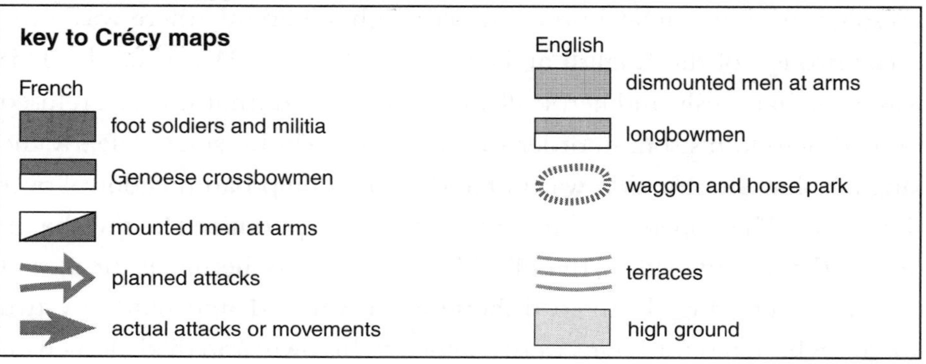
key to Crécy maps
French
foot soldiers and militia
Genoese crossbowmen
mounted men at arms
planned attacks
actual attacks or movements
English
dismounted men at arms
longbowmen
waggon and horse park
terraces
high ground

The Battle of Crécy, 26 August 1346

The English under King Edward III
The French under King Philip VI
English forces: 8–14,000 (2–3,000 men-at-arms, 5–10,000 archers, 1,000 spearmen)
French forces: 20–25,000 (including 12,000 men-at-arms and 6,000 crossbowmen)
English casualties: unknown but slight
French casualties: considerable

Critical Moments

Edward's forces deploy towards the high ground
Edward uses the horse lines and baggage park to protect his rear
The English archers outshoot the Genoese crossbowmen
The French knights charge to disaster
The English men-at-arms overwhelm the French

rode them down in their eagerness to get at the English, who 'continued shooting as vigorously and quickly as before; some of their arrows fell among the horsemen, who were sumptuously equipped, and killing and wounding many, made them caper and fall among the Genoese, so that they were in such confusion that they could never rally again.' This is when the men-at-arms discovered why the crossbowmen had faltered. According to the contemporary chronicler Jean Le Bel, 'the archers shot so fiercely that those on horseback suffered from these deadly barbed arrows: here, one horse was refusing to go forward, there, another leaping about as if maddened, here, was one bucking hideously, there, another turning its haunches to the enemy.' Those cavalry who did reach the English men-at-arms had lost all sort of order and hence any possible impact. But it was the English men-at-arms, not only the archers, who won the day for Edward. Hand to hand fighting was necessary, and Geoffrey Le Baker writes that, 'When fighting with the English men-at-arms, the French were beaten down by axes, lances, and swords. And in the middle of the French army many were crushed to death in the press without being wounded.' Many and uncoordinated charges were simply no solution to the tactical problem, although apparently there was some breakthrough of the English archers of the Prince of Wales' battle. This was when, famously and heroically, Froissart records that Edward refused to send help to his son, in order that he might win his spurs! Clearly the king felt that the situation was in hand, while Philip had no control over the action. The hopeless nature of the contest is epitomized by the charge on the French side led by the blind King John of Bohemia at the end of the day. Preferring death to dishonour, he was led into battle by two knights whose mounts' reins were chained to his own, and died along with them in the press. The Prince of Wales took the dead king's motto 'Ich Dien' (I Serve) for his own.

The Results of the Battle

Crécy proved the superiority of the English tactical system. 1,542 French knights, recorded by heralds, lay dead in front of the Prince of Wales' 'battle alone'. Many hundreds of others died on that day, or like some late-arriving infantry were slaughtered the day after. The defeat was a total humiliation for King Philip and French chivalry. But was Edward lucky, or had he planned the battle long ago? In later centuries it was common for generals to identify likely battle sites and seek to fight on them, as both Frederick the Great and Wellington certainly did. Edward was certainly experienced enough to have planned something similar, and Crécy was a perfect site for a defensive battle in an area well known to him. By combining his men-at-arms and archers with a strong position he had achieved a tactical surprise against a much stronger enemy who did know what to expect, but was drawn by force of circumstances into a losing encounter. This was due to generalship of the highest order displayed by Edward.

The longer term results of the battle were indeed significant. Edward was able to proceed with a siege of Calais, the well-defended port on the coast of France nearest to England. This took a full year to capture; but the memory of the defeat at Crécy meant that Philip's relief attempts in October 1346, and again in May and July 1347 proved futile. He simply dared not risk open battle again, and the defenders of Calais were abandoned to their fate. Philip was paying the penalty for a combination of arrogance, rashness, and weakness in submitting to the demands of his young nobles who were over-confident of victory. With hindsight, it is difficult to see how the French could have defeated Edward on that stormy August afternoon. Their troops lacked the flexibility and expertise to overcome an enemy confident in his arms and tactics. Perhaps Froissart should be allowed the last word when he says: 'That day the English archers gave a tremendous advantage to their side. Many say that it was by their shooting that the day was won.'

MATTHEW BENNETT

Further Reading

Bradbury, J., *The Medieval Archer* (Woodbridge, 1985)
Burne, A.H., *The Crécy War* (London, 1955)
DeVries, K., *Infantry Warfare in the Early Fourteenth Century* (Woodbridge, 1996)
De Wailly, H., *Crécy 1346: Anatomy of a Battle* (Poole, 1987)

Cambrai

20 November–7 December 1917

The New Invention

'When we… looked across towards the positions now held by the Germans, there was none that could not help thinking of the great and successful work of our troops during those four splendid days from 20th November onwards, of the ground that had been captured from the enemy, of the comrades who gave their lives for that ground. And now that territory no longer belonged to us, but was once again in the hands of the foe.' From *The Royal Tank Corps Journal, 1927*

The Western Front in World War I has become notorious as an illustration of failed generalship and poor planning. It is often forgotten that eventually the problems of how to break through the enemy trench defences were solved, enabling both sides to make dramatic advances in 1918. Many factors contributed to this, but among the most important were a number of tactical and technological innovations coming over half-way through the war, which allowed commanders to plan to surprise the enemy. On 21 November 1917, church bells were rung in England for the first time since the beginning of the war, to celebrate the British success in the Battle of Cambrai, one of the first small tests of these new innovations. Although the battle ended in the first week in December with few of the initial British gains remaining, the first day of the offensive showed the Allies what could be achieved when an emphasis was placed upon the attainment of surprise and the pursuance of all arms cooperation.

The Origins of the Battle

The origins of the Battle of Cambrai lay in the situation in which the British found themselves after the failure of the Third Battle of Ypres (or 'Passchendaele') in early November 1917. Although it was late in the campaigning year, the commander in chief of the British Expeditionary Force, Field Marshal Sir Douglas Haig, was desperate for a victory with which to appease the braying politicians, encourage the increasingly war-weary civilians at home, and boost the morale of the troops at the front. This is not to say that the battle was devised merely in order to protect Haig at a difficult time in the war for him. Although a victory anywhere would have helped, after a year that had started so brightly with the British and Canadian success at the Battle of Arras in April had faded in the autumnal gloom of Ypres, a breakthrough at Cambrai was a distinct possibility in late 1917.

The Cambrai sector of front was chosen for the attack as it was well suited to British offensive requirements, especially those which were to put so much emphasis upon the attainment of surprise and the use of large numbers of tanks. Despite its name, the battle was fought in the open fields and villages south of the town of Cambrai itself, which was a major communications centre just six miles behind the German front line. The countryside consisted of rolling fields and gentle valleys; the ground was hard chalk which had not been pitted by constant shelling, and there were plenty of woods and ruined villages in the vicinity which would prove useful for the concealment of tanks, infantry, and guns prior to the assault. Add to this the major advantage that the Cambrai sector was being treated as rest area for exhausted German troops (which meant that the six-mile front to be attacked by the British contained only two German divisions and 34 guns), and this part of the front looked the ideal place to conduct a British attack.

Clearly Cambrai offered many offensive temptations, but the British never forgot that the strength of the German defensive positions in the area negated many of their advantages. Along their front the Germans had built a series of strong defensive positions which they regarded as impregnable, and which the British called the Hindenburg Line (after the German Chief of the Great General Staff, Field Marshal Paul von Hindenburg). The Hindenburg Line position in front of Cambrai was five and a half miles deep, made excellent use of the terrain and incorporated thick belts of barbed wire, machine gun nests, and trenches designed to stop the progress of tanks. British success at Cambrai, as a result, was far from guaranteed.

The British Plan

The British plan, devised by General Sir Julian Byng's Third Army, sought not only to break into the German front line, but also to achieve full-scale penetration and exploitation. The first Third Army objective was the main Hindenburg Line between the position known to the British as Bleak House and the Canal Du Nord, the second objective was the Hindenburg Support Line, and the third objective was exploitation in the direction of Cambrai. This plan was passed by Haig who, despite a preference for more limited objectives, and fears about the lack of troops available for the attack, decided not to alter the scheme. The commander in chief did, however, reserve the right to curb the offensive after 48 hours if it showed no signs of success.

To achieve Byng's objectives, the Cambrai plan had to be bold and highly innovative. One of the boldest moves was the way in which tanks were to be used; indeed, the employment of the tank and the Battle of Cambrai have become synonymous in the minds of many who think about the use of armour in World War I. Not only were tanks used in the battle in numbers that had never before been seen, but the plan was based around what these machines could achieve. The Battle of Cambrai plan shows that the tank was no longer viewed as an appendage that might be of use in some way; it reveals that by November 1917 many, including Haig, regarded the tank as a key offensive weapon. Such thinking did not disregard the weaknesses that the early tanks had in abundance, such as a proneness to break down, slowness, and lack of mobility. But during the previous year the introduction of an improved motor, extended range, more effective fire power, and strong frontal armour had at least ameliorated some of the most obvious frailties. Thus, by the end of 1917, it was thought that if hundreds of machines were to be used in an attack on level ground that had not been cratered by artillery fire or soaked by rain, armour was capable of creating its first major impact upon the battlefield.

However, the British planners of the Battle of Cambrai did not only rely upon tanks for success. As important as the use of massed armour were the new tactics that had been devised for a much older weapon of war, the artillery. Over the course of 1917, the artillery had developed tactics that managed to engage enemy trenches and the enemy's guns without 'ranging fire', which in the past had been necessary in order to discover exactly where the shells from the guns would land. The import-ance of this 'silent registration' (so called because it removed the need to fire ranging shells in order to 'register' targets before an offensive, and so give away both the positions of the guns and the intention to attack) was enormous, as it allowed for operational surprise to once again be achieved. This development had come about as the result of lessons learned from previous battles: the application of technology that led to more accurate maps, survey techniques, and meteorological data, better aerial photographs (which provided more detailed knowledge about the layout of enemy defences) and the ability to fix enemy gun positions by sound ranging and flash spotting.

What these new artillery tactics meant in terms of the British plan was that, rather than there being days of preliminary bombardment, the guns would not open up until Zero Hour (the hour of the attack) on the first day of the offensive. The roles previously carried out by the preliminary bombardment still had to be accomplished, but in order to gain surprise they were conducted in a different manner. Instead of days of bombardment destroying the enemy barbed wire, the tanks would crush it. The German trenches would be destroyed by a 'jumping' barrage which would move in a series of fairly big 'lifts', each of a few hundred yards onto the next German position, and counterbattery work against enemy artillery would only begin once tanks and infantry had started to move forward.

Although the replacement at Cambrai of the preliminary bombardment with new artillery techniques meant that fewer shells would be fired at the enemy, it was hoped that when greater precision was combined with the achievement of surprise, the artillery would actually become far more effective. The achievement of such precision and surprise were eventually central to the British success on the first day of the Battle of Cambrai.

The British Attack

The British offensive at Cambrai began at 6.20 a.m. on 20 November, when over 300 fighting tanks led six infantry divisions of Third Army's III Corps and IV Corps into the attack. As the infantry and tanks moved off towards their objectives, the all-important artillery barrage began from the 1,000 guns that had thus far lain silent. A mixture of smoke, high explosive, and shrapnel rained down upon German positions with an accuracy that not only destroyed large parts of their defences, but also moved forward in a manner that forced the enemy to stay in their deep 'dugouts', protected shelters dug beneath the trench defences.

The surprise of the attack that morning, together with successful cooperation between the infantry, armour, artillery, and the fourteen squadrons of the Royal Flying Corps that flew overhead, to observe as well as to harass the German ground forces, meant that early success was achieved. The tanks rolled forward followed by columns of infantry through what the Germans had thought was impregnable barbed wire. When the tanks reached the German trenches, designed to be too deep and broad for armour to cross, the machines were able to drop fascines and move over them. Each fascine was 3 to 4 feet in diameter and consisted of 60 bundles of brushwood bound together by a chain and weighing about $1\frac{1}{2}$ tons.

This technology worked well and enabled the British to maintain their forward momentum. One tank commander later recalled that 'Just before half-past six the barrage commenced ... and we started off. Our first bump came fairly soon. We climbed a bank, crashed through the hedge on top and came down heavily on the other side. Our tank weighed some 28 tons. When it lurched it threw its crew about like so many peanuts ... We clattered across No-Man's Land, crushing a path for the infantry through 50

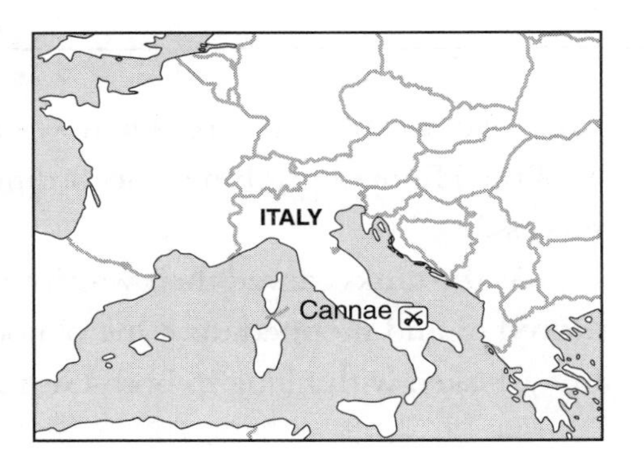

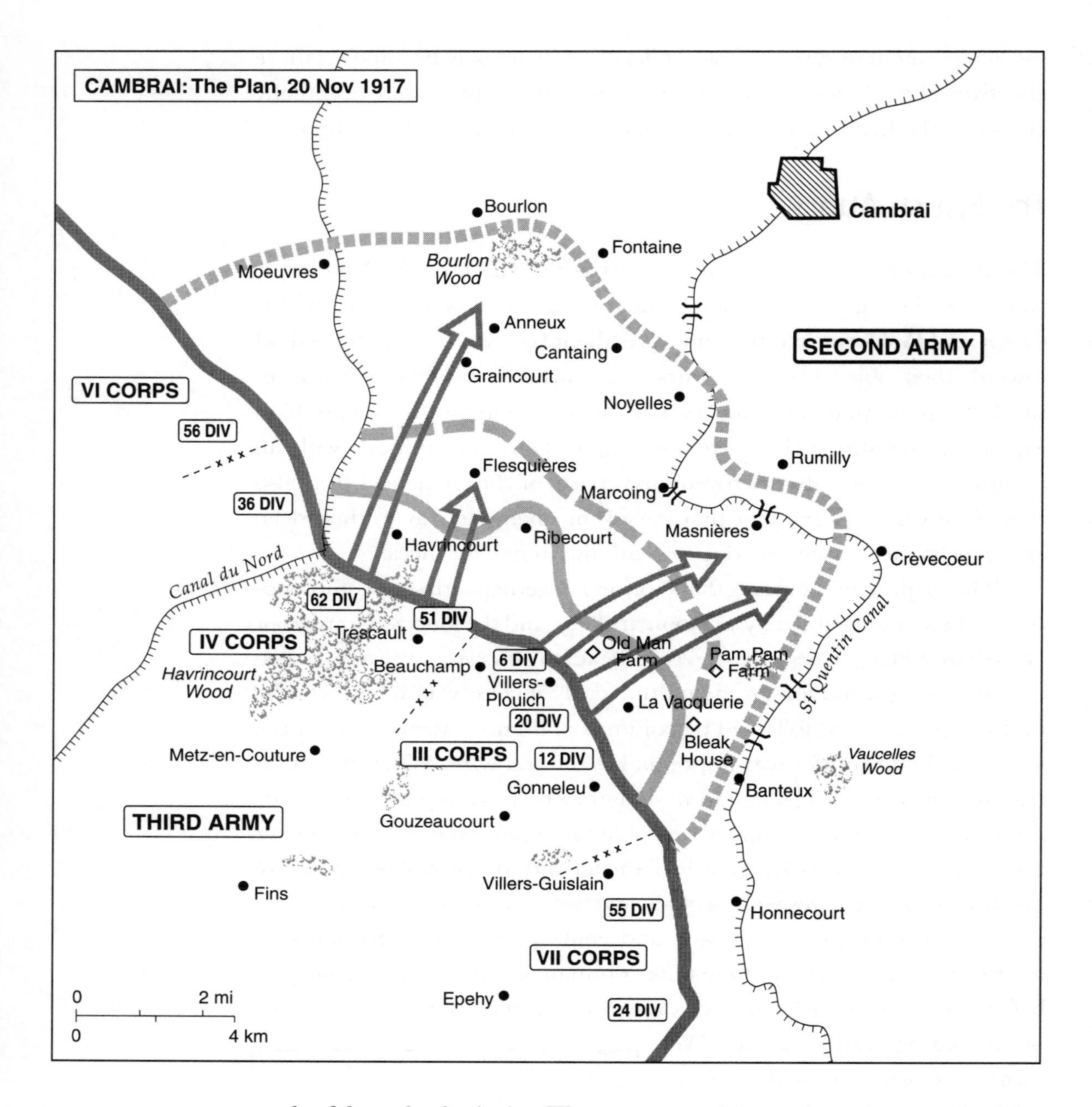

yards of dense barbed wire. Then we crossed the main and reserve trenches of the Hindenburg Line – according to plan. The fascines were a great success.'

As the tanks carved their way into the German positions, the infantry moved in and mopped up. One platoon commander, Lieutenant John F Lucy, tasked with filling in both German and British trenches so that the

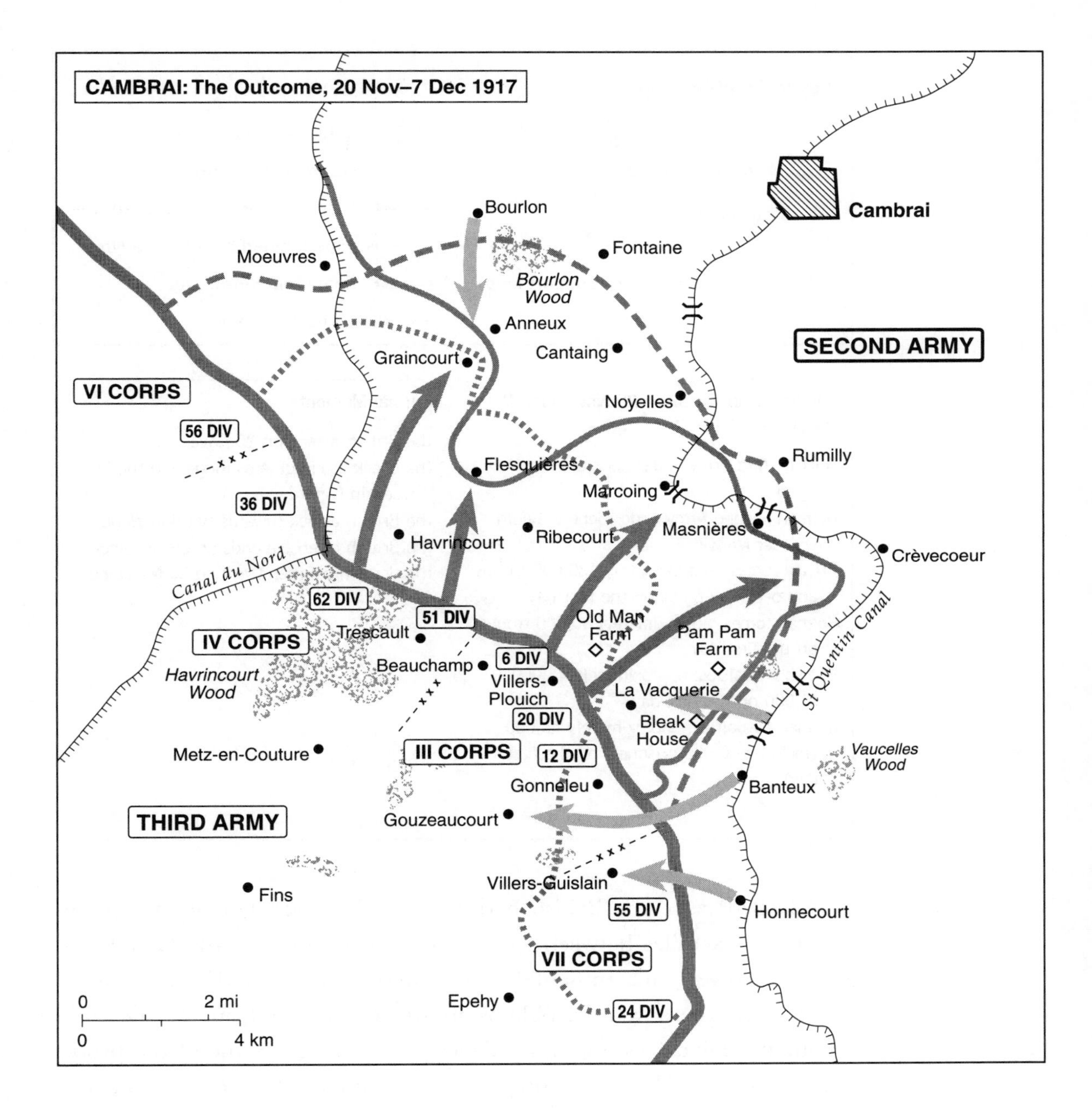

artillery could get forward and continue their support, remembered that the first few hours of the Battle of Cambrai were 'astounding' as he watched 'whole sections, even platoons, of infantry strolling up the opposite enemy slopes behind [the tanks]'. This was a far cry from previous offensives, when the infantry often found the German wire uncut and machine gunners ready to cut them down.

key to Cambrai maps

original Front Line 20 Nov
planned Allied attacks
Allied objectives
1st
2nd (Hindenburg Support Line)
3rd
actual Allied attacks
German counterattacks
Front Line on evening of 20 Nov
limit attained by Allies 29 Nov
final Front Line 7 Dec
- x x x - corps boundary

The Battle of Cambrai, 20 November–7 December 1917

British Third Army under General Sir Julian Byng
German Second Army under General Georg von der Marwitz
British forces: approximately 120,000 troops and over 300 tanks on the first day
German forces: approximately 40,000 troops on the first day
British casualties: approximately 45,000 (4,000 on the first day)
German casualties: approximately 45,000 (including 6,000 prisoners on the first day)

Critical Moments

The British attack on 20 November
The check at Flesquiéres Ridge and the St Quentin Canal
The British attack towards Bourlon Wood
The British offensive ends on 29 November
The German counterattack on 30 November

By 8.00 a.m. the British had overrun the main Hindenburg Line along the six miles between the St Quentin Canal and the Canal Du Nord. However, while most of the German support line had been taken by 11.30 a.m., there were problems in the centre, in front of the 51st (Highland) Division's objective of Flesquiéres Ridge. Some of the problems encountered by the infantry here came about as a direct consequence of a number of important alterations made by the divisional commander to the tank–infantry cooperation tactics that were being used by all other divisions. With the infantry struggling to move forward, the tanks continued their advance unsupported. Thus, when the slow-moving tanks came in range of a number of German guns behind the ridge that had not been destroyed by the British artillery, they suffered heavy losses.

The difficulties at Flesquiéres reflected a far greater problem.

Although by the early afternoon of 20 November the British had achieved a massive surprise, and had managed to advance to a depth of 9,000 yards (a penetration not accomplished in four months of fighting during the Third Battle of Ypres), the attack had failed to achieve the all-important breakthrough. Although the failure to attain this on the first day was not devastating in itself, it was understood by the British Higher Command that with manpower resources for the offensive so severely restricted, a breakthrough would have to be achieved very quickly. The 4,000 British casualties and 179 tanks lost on the first day did little to lessen the resources problem. Although reinforcements were found to help bolster numbers after 20 November, there was no possibility of so many tanks being replaced quickly and, to add to operational problems, the cost was bound to increase now that the element of surprise had been lost.

Nonetheless, despite these problems, a decision was taken to continue the offensive in the hope, having weakened the Germans significantly on the first day, that they would collapse after just one more push. Thus, although in late November 1917 new offensive ideas took an important step forward, more traditional assumptions and reactions took hold when the innovative methods failed to realize decisive results early enough. This desire to keep plugging away at the Germans (whilst guarding against overcommitment of troops) led Haig to order the end of the attack on his right, whilst continuing the push towards the Bourlon Ridge on his left. Meanwhile, however, the Germans flooded the Cambrai area with reinforcements of such numbers that any further British attack would be rendered impotent.

Having survived the initial British assault, the Germans were well placed to defend against subsequent British offensive efforts during the final week of November. Although the Germans abandoned Flesquiéres during the night of 20–21 November, British progress towards Bourlon Ridge was disappointing and resulted in heavy casualties. After the innovation of the first day of the battle, many of the old Western Front problems began to seep back to haunt commanders as the days passed. The exploitation of any success, for example, remained a difficulty, as the artillery still lacked the sort of mobility that would allow it to be moved forward quickly to support subsequent attacks. Poor communication remained a problem, and keeping troops supplied with all that they needed when they left their own front line was nearly impossible. Gradually, as the early coordination between guns, tanks, and infantry steadily waned, a less sophisticated slogging match ensued.

On 27 November 1917, with the British having been fought to a virtual standstill and casualties mounting, Haig decided to call a halt to the offensive, which came to an end two days later. The problem was, however, that the attack had created a large salient in which troops and artillery were vulnerable to enfilade (or flanking) fire from either side, and constantly under threat of being cut off.

The German Counterattack and the End of the Battle

The obvious thing for the British to do was to withdraw from the salient. But as they were preparing to do this, the enemy counterattacked on 30 November, opening with an unregistered 'hurricane' bombardment of great intensity. Although the Germans attacked the northern part of the salient with little comparable success owing to British concentrations there, their main blow was aimed at the southern end of the British line around Banteux, where the British were at their weakest. This time it was the British turn to be surprised. German infantry tactics sought to exploit any fragility in the British line, and after an intense bombardment for over an hour, the infantry vanguard infiltrated through areas weakened by the artillery, by-passed strong points, and then dealt with the British guns to the rear. By 10.30 a.m, the Germans had managed to advance eight miles into the southern British sector, which placed them well beyond what had been the original British line before the 20 November offensive.

Complete disaster for the British was averted by the timely arrival of reinforcements, and the increasing problems that the Germans also had in sustaining their push. As German momentum petered out, the British began to stabilize their line and fight back. With no chance of severing the base of the salient remaining, the exhausted Germans brought their counterattack to an end on 3 December. The battle itself finally came to an end four days later after one final British withdrawal to a more suitable defensive position. By the time it ended, the battle had involved almost 20 British divisions, a quarter of their forces on the Western Front.

The Battle of Cambrai saw startling British success quickly followed by setback and was, as a consequence, a draw. The gains achieved on 20 November were quickly wiped out by the German counterattack, which managed to inflict heavy losses upon the British and severely undermined their newly bolstered morale. For the British the battle was most certainly not the success that they had wished. They had retained a little more ground of tactical importance in the northern part of the sector than they had relinquished in the southern part, but this was not the sort of achievement for which they had hoped. The battle cost the British 45,000 casualties, at least as many as those sustained by the enemy, and two-thirds of their tanks.

Nevertheless, the Battle of Cambrai did reveal that fighting on the Western Front had entered a period of transition. Many of the new fighting methods that were to be on show in the final year of the war were previewed in November 1917. The German hurricane bombardment and use of infantry infiltration was seen in their spring 1918 offensives, while the employment of an unpredicted bombardment and the coordin-ation of the artillery, infantry, and tanks were all to play key roles in the British offensives that ended in the armistice in November 1918. However, as clearly visible as the new techniques on show by both sides during the Battle of Cambrai was the chronic problem of how to turn tactical success into operational victory. It was to be another eight months before the conditions were right for the British to find a solution to this problem.

LLOYD CLARK

Further Reading

Griffith, P. (ed.), *British Fighting Methods in the Great War* (London, 1996)
Smithers, A.J., *Cambrai: The First Great Tank Battle* (London, 1992)
Travers, T., *How the War was Won: Command and Technology in the British Army on the Western Front, 1917–18* (London, 1992)
Wilson, T., *The Myriad Faces of War: Britain and the Great War, 1914–1918* (London, 1986)

Blitzkrieg

10 May–22 June 1940

The Shock of Surprise

'The highest French leadership either could not or would not grasp the significance of the tank in mobile warfare.' Heinz Guderian

In some battles, if only by the laws of chance, almost everything goes right for one side and wrong for the other. When this happens, the shock is so overwhelming that it produces not just defeat, but the disintegration of most of the defeated army into a leaderless mob. But if chance plays a part in this, better plans and better generalship are also an important contribution, so that no matter how bravely the enemy fights back their forces stand little chance. A particularly good ex-ample of this is the German campaign of spring 1940, popularly known as the 'Blitzkrieg' (literally, 'lightning war'), which overran Belgium and the Netherlands, reduced France to a rump satellite state of the greater German Reich, and ejected British forces from the continent, all in a matter of weeks. This stunning victory was achieved by armed forces that enjoyed no decisive advantage in terms of numbers or technology. Allied defeat resulted from differences in how the two sides conducted war, and the way that the German plan exploited major defects in the Allied plan.

The Opposing Armies

Memories of World War I played a major part in determining the way that the Allies attempted to fight in 1940. Britain refused to pull its weight in continental land warfare, committing only ten infantry div-isions to the continent. France also entered the war reluctantly, and many historians detect the seeds of the French military collapse in its pre-war political divisions. However, reluctance to fight another great European war was common to all belligerents, and a more detrimental legacy of 1914–18 was its impact on how the Allies wished to conduct the war.

The Allied governments believed that the Germans had missed their opportunity for a quick victory in 1939, and this optimism was based on some powerful military assets. Numerically the Allies matched the Germans in troops, and they outnumbered them in artillery by 5,600 guns, chiefly the excellent French artillery. Along the entire length of the Franco-German frontier a powerful belt of fortifications – the Maginot Line (named after French Minister of War André Maginot) – rendered any direct German attack on the Allies problematic. Greatly superior Allied naval power ensured that Germany was blockaded from much of the outside world, enhancing the ability of the Allied economies to out-produce Germany. In a long war, which the Allies believed they could force Germany to fight, the military balance would move in their favour. This 'Maginot Mentality' was not as unrealistic as some histor-ians suggest, but most of the Allied leaders failed to appreciate that mechanized forces and air power gave Germany the potential to wage a swift and decisive campaign.

However, the Allies were also capable of waging mobile warfare. Indeed they possessed more tanks and vehicles than the Germans. But most Allied mechanized assets were in what the British called 'penny-packets' (scattered in small groups along the Allied front). The French had also formed three armoured divisions, which were powerful all-arms formations based on modern tanks, and were forming a fourth; the British had one armoured division which fought in the later part of the battle, and the infantry divisions of the British Expeditionary Force (BEF) were lavishly equipped with motor vehicles. However, Allied armour formations were 'tank heavy' and failed to coordinate well with other arms. Allied tanks tended either to resemble a mobile version of the 'pill-boxes' intended for static defence, or were designed to deliver an updated version of the cavalry charge. For instance, the French B-1 tank was armed with a 75-mm gun in its hull and a 47-mm gun in its turret – no other tank in 1940 packed such a punch – but it was slow, had a limited range and most of them had no radios, which were indispensable for coordinating with other arms.

Although the Allies were aware of the value of mechanization, in 1940 the Germans had the advantage. German mechanized assets were concentrated into ten armoured divisions, known as Panzer divisions from *Panzerkampfwagon* (PzKw) (the German for tank). These combined significant numbers of motorized infantry, artillery, engineers, and signallers with their tanks. The *PzKw I* and *PzKw II*, the most numerous German tanks, were very inferior to those of the Allies. However, the best German tanks, the *PzKw III, PzKw IV*, and the Czech-manufactured *PzKw 35/38t*, all possessed a gun that enabled them to engage most Allied tanks with some prospect of success, they were protected by adequate armour, and robust engines gave them a good range and an acceptable speed. German tanks had all-round qualities, and were better suited for a role in an all-arms mechanized formation.

Germany had no clear numerical or technological advantage on the ground, but in the air the Luftwaffe (the German airforce) substantially outnumbered the Allied air forces, many of whose planes were obsolete. Unlike the Allied air forces, especially the British Royal Air Force (RAF), the Luftwaffe was closely tied to German ground forces and close air support was integral to the way in which Germany conducted ground operations.

The Rival Plans

During the inter-war years German military thinkers focused on winning wars quickly, and this desire for speed was enhanced by the Nazis' lack of faith in the German people's stamina for a long war after the experience of World War I. Many German generals doubted that one campaign could defeat the Allies, but in 1940 commanders who believed that a combination of mechanization, airpower, and boldness could deliver rapid results were given the opportunity to vindicate their faith.

The Allied plan of 1940 had its origin in Belgium's decision in 1936 to abandon its military alliance with France. This deprived the Allies of the chance of meeting a German attack on the powerfully fortified, and relatively short, Belgian–German frontier. It also ensured that the Belgian Army faced the initial German attack alone, and it was so badly mauled that it made a negligible contribution to the defence of its country. The Allies, expecting a German plan which would attempt to outflank the Maginot Line to the north, could have abandoned any plan to defend neutral Belgium, meeting the German offensive on the Franco-Belgian frontier. However, the French Supreme Commander, General Maurice Gamelin, advocated an advance into Belgium to block the anticipated German attack. In November 1939, Gamelin persuaded the Allied governments to adopt his 'Plan D', named after the River Dyle in Belgium, which was the line that he hoped to hold, a swift advance into the Low Countries in response to any German attack upon them, as far as the Dyle and the Dutch city of Breda in the north.

This planned Allied response to the anticipated German invasion of the Low Countries was fatally flawed. Putting the strongest and most mechanized Allied forces opposite the western Belgian frontier meant that only a modest number of third-rate French divisions covered the Namur–Sedan gap, the hinge between this northern wheel and the Maginot line, in the region of the Ardennes forest. The hilly and tree-covered Ardennes was deemed easily defensible, and unsuitable terrain for mechanized warfare. Gamelin believed that any German thrust through the Ardennes would take time to prepare, and would be slowed by Belgian defences. The advance into the Low Countries depleted Allied reserves, leaving them with little strength to deal with the unexpected. Many Allied commanders feared that an advance into the Low Countries would place them at a disadvantage, but Gamelin does not seem to have considered the problems of retreating from the Low Countries if worsted by German forces.

In contrast, when the final elements of the Polish army surrendered on 1 October 1939, Adolf Hitler ordered an offensive against France before the end of the year. On 19 October, the German High Command – *Oberkommando der Wehrmacht* or OKW – issued the plan *Fall Gelb* (Case Yellow), envisaging an attack through the Low Countries with the main weight of the offensive north of the Ardennes. This was the

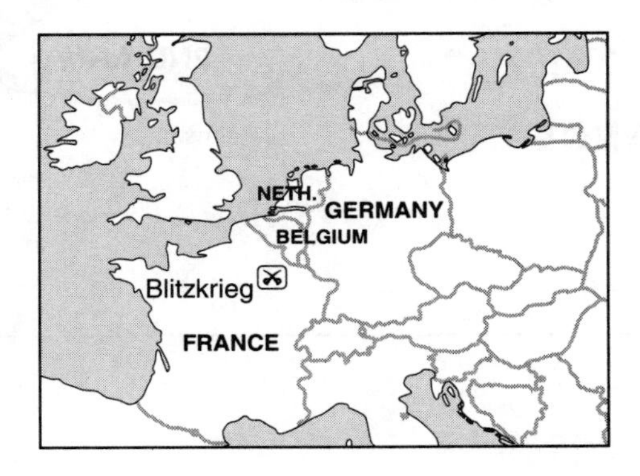

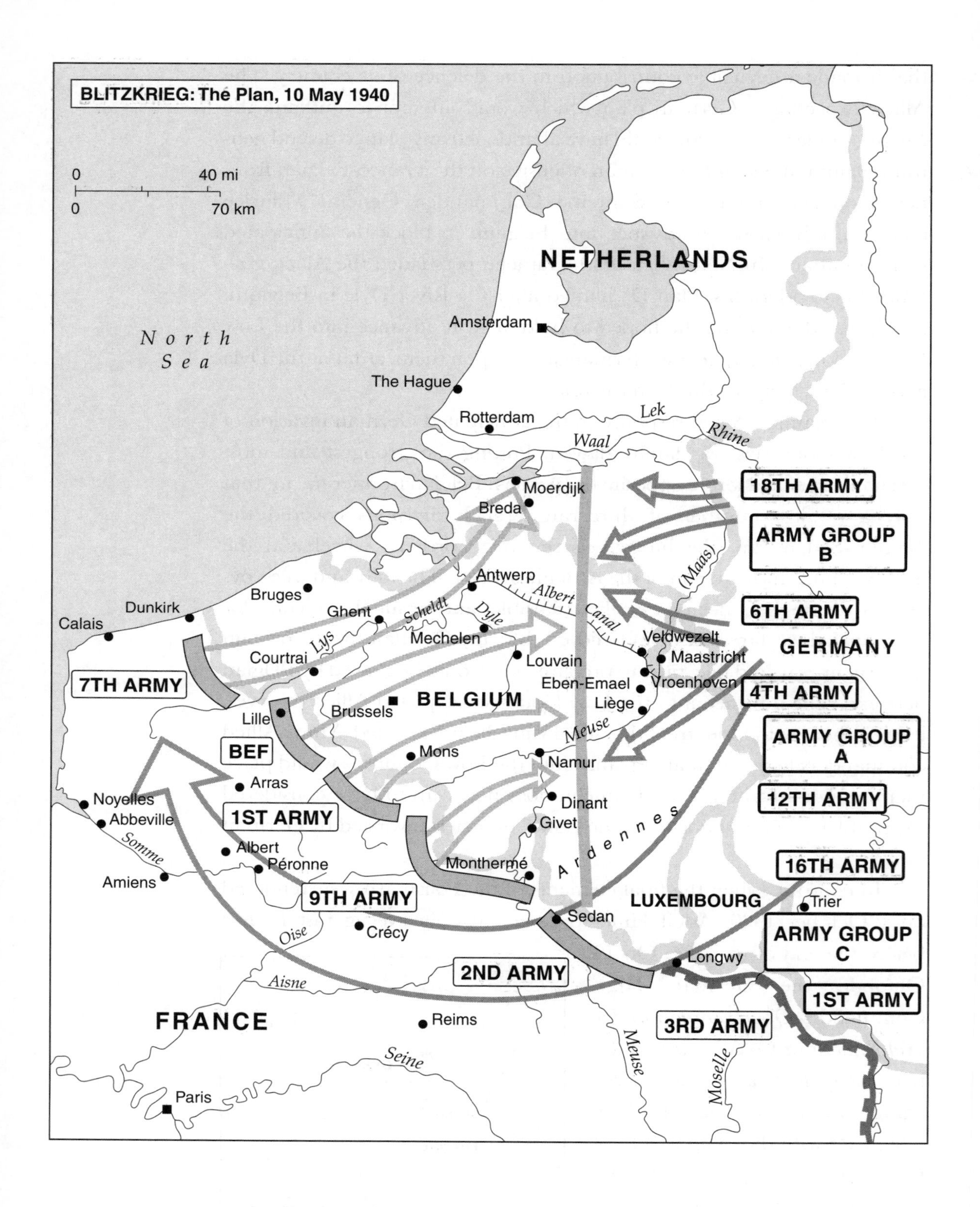

BLITZKRIEG: The Plan, 10 May 1940
0
40 mi
0
70 km
NETHERLANDS
North Sea
Amsterdam
The Hague
Rotterdam
Lek
Waal
Rhine
Moerdijk
Breda
18TH ARMY
ARMY GROUP B
(Maas)
Antwerp
Bruges
Albert
Canal
Dunkirk
Ghent
Scheldt
Dyle
6TH ARMY
Calais
Lys
Mechelen
Veldwezelt
Maastricht
GERMANY
Courtrai
Louvain
Eben-Emael
Vroenhoven
7TH ARMY
BELGIUM
Liège
4TH ARMY
Lille
Brussels
Meuse
ARMY GROUP A
BEF
Mons
Namur
Arras
Dinant
12TH ARMY
Noyelles
1ST ARMY
Abbeville
Givet
Ardennes
Somme
Albert
Péronne
Monthermé
16TH ARMY
Amiens
9TH ARMY
LUXEMBOURG
Trier
Sedan
Oise
Crécy
ARMY GROUP C
Longwy
2ND ARMY
Aisne
1ST ARMY
FRANCE
Reims
3RD ARMY
Meuse
Moselle
Seine
Paris

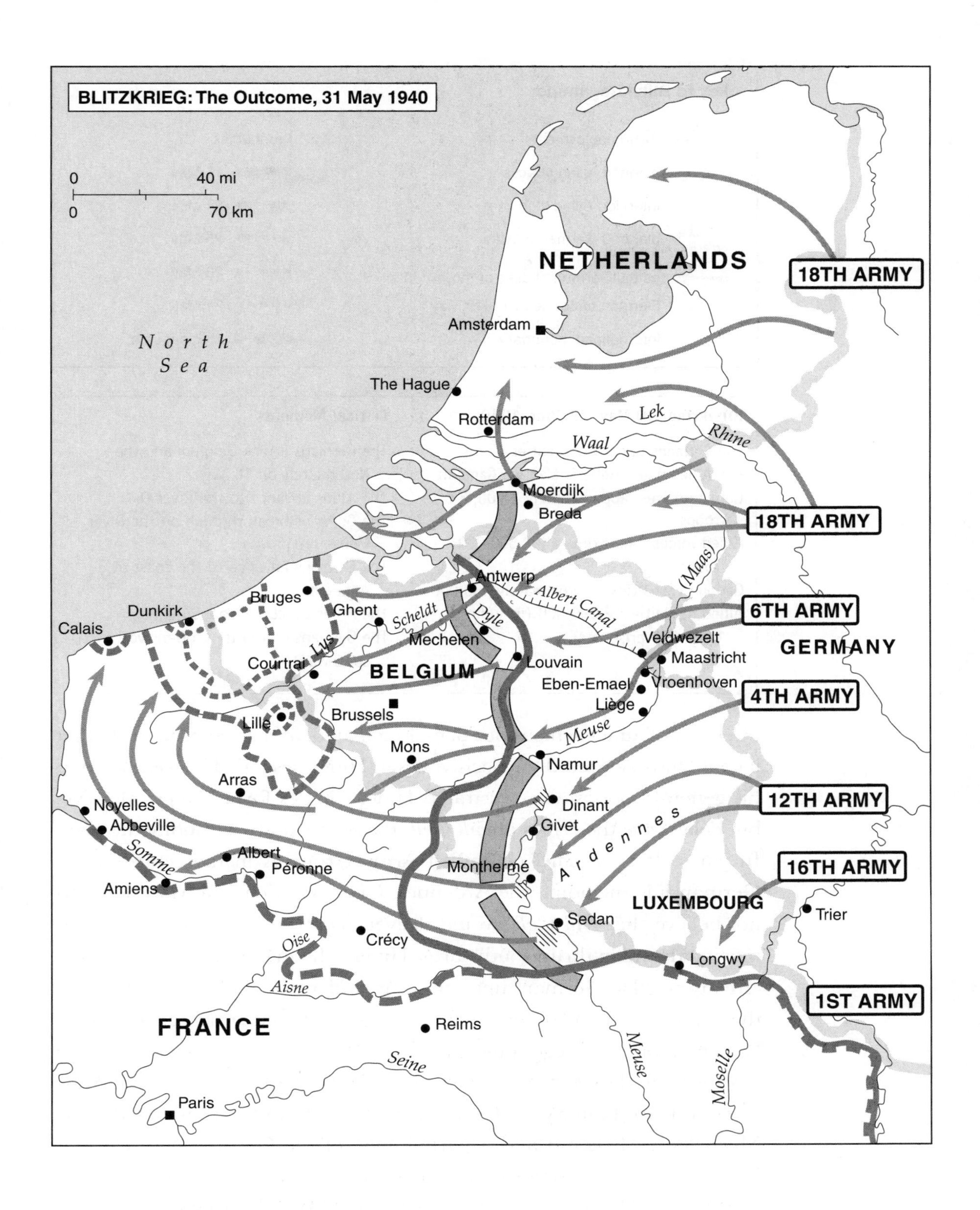
BLITZKRIEG: The Outcome, 31 May 1940
0
40 mi
0
70 km
North Sea
NETHERLANDS
18TH ARMY
Amsterdam
The Hague
Rotterdam
Lek
Rhine
Waal
Moerdijk
Breda
18TH ARMY
(Maas)
Antwerp
Albert Canal
Bruges
Ghent
Scheldt
Dyle
Dunkirk
Calais
Mechelen
6TH ARMY
Veldwezelt
Maastricht
GERMANY
Lys
Courtrai
BELGIUM
Louvain
Eben-Emael
Vroenhoven
4TH ARMY
Liège
Brussels
Lille
Meuse
Mons
Namur
Arras
Dinant
12TH ARMY
Noyelles
Abbeville
Givet
Ardennes
Somme
Albert
Péronne
Monthermé
16TH ARMY
Amiens
LUXEMBOURG
Trier
Sedan
Oise
Crécy
Longwy
Aisne
1ST ARMY
FRANCE
Reims
Seine
Meuse
Moselle
Paris

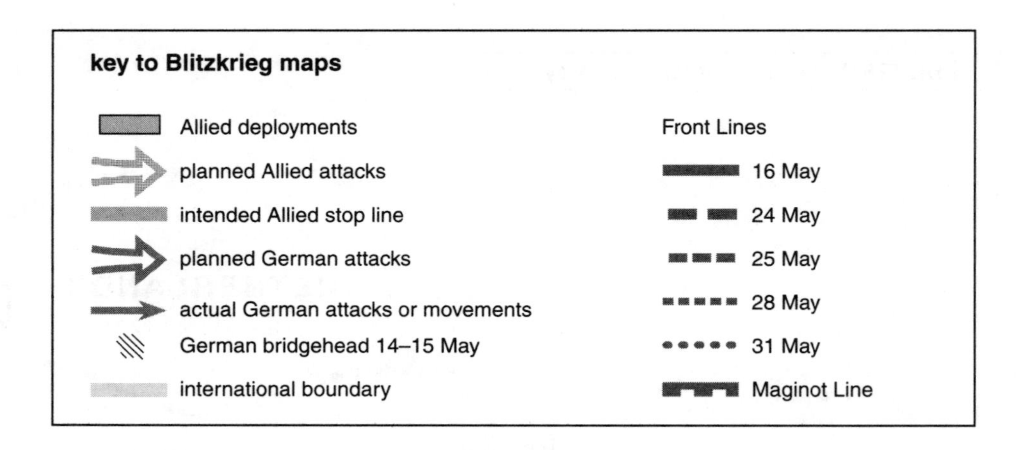

Blitzkrieg, 10 May–22 June 1940

The Germans under Adolf Hitler
The Allies under General Maurice Gamelin
German forces: approximately 2,200,000 troops
Allied forces: approximately 2,500,000 troops
German casualties: approximately 150,000
Allied casualties: 2,390,000 (including 1,900,000 prisoners)

Critical Moments

The Germans attack Belgium and the Netherlands on 10 May
The Allies advance to the River Dyle
The Germans break through on the River Meuse on 13 May
The Germans advance to the coast at Abbeville
The Allies evacuate from Dunkirk
The Germans review their offensive southward

direction of attack which the Allies anticipated, and it seems unlikely that it could have delivered the decisive result Hitler desired. The weather, and his generals' opposition, frustrated Hitler's hopes for a winter offensive, but delay gave the critics of *Fall Gelb* the opportunity to modify the plan. In particular, the chief of staff of Army Group A in the centre of the German line including the Ardennes, General Erich von Manstein, did not believe that *Fall Gelb* would deliver decisive results, and advocated moving the main thrust southwards. Hitler, who in practice if not in name already acted as German supreme commander, also had his doubts about the plan, and in October 1939 he suggested that a successful attack launched south of Liège could cut off the Allied armies of the northeast by advancing to the coast. Modifications of *Fall Gelb* were already underway, when on 10 January, a German courier aircraft crash-landed near Mechelen in Belgium and a partial copy of the German plan fell into Belgian hands. This incident increased the impetus for major changes; on 17 February Hitler met von Manstein, and, finding his views echoed by a military professional, decided to impose a new plan on his generals.

The revised plan for *Fall Gelb* was very bold, so much so that throughout the offensive significant elements of OKW feared that its boldness could backfire. Army Group C, the weakest of Germany's three army groups opposing the Allies with only 17 divisions, was deployed along the German–French frontier. Its task was to ostentatiously prepare to assault the Maginot Line, and to leak false plans for an outflanking movement through Switzerland. Army Group B with over 26 divisions and 3 Panzer divisions was deployed opposite the Dutch and northern Belgian frontiers. Its task was to invade the Low Countries with massive Luftwaffe support, enticing the Allied armies to advance rapidly into Belgium and preventing their rapid withdrawal in good order. Because it was important for Army Group B to advance quickly, it was assigned most of Germany's scarce airborne forces, and in the initial stages of the offensive the support of most of the Luftwaffe. Army Group A, with over 37 divisions and 7 Panzer divisions, was deployed opposite the southern Belgian and Luxembourg frontiers. It was to advance swiftly but stealthily through the Ardennes, cross the River Meuse, and then if all went well drive northwest to the coast to cut off the Allied forces in the Low Countries. Provided that operations to the north went according to plan, Army Group A could expect the support of the bulk of the Luftwaffe once it commenced its attempts to cross the Meuse.

The German Attack

The German offensive commenced at 5.35 a.m. on 10 May. The Luftwaffe speedily established its superiority, attacking Allied communications, formations on the move, and airfields, and all too often catching Allied planes on the ground. Army Group B advanced swiftly into the Low Countries, aided by airborne and raiding forces. The powerful Belgian fort of Eben-Emael was taken by one of these small raiding forces, the gliders of the Koch Assault Detachment landing on its flat roof. Elsewhere in Belgium and the Netherlands, German airborne forces seized intact bridges over the rivers Waal, Lek, and Maas, and the Albert Canal. The Dutch were unable to recapture the bridge at Moerdijk over the Waal, and on the 13 May advanced elements of 9th Panzer Division were crossing into 'Fortress Holland'. In just four days the Dutch were totally defeated, the bombing of Rotterdam on 14 May hastening the Dutch formal surrender. To the south, the Belgians were faring little better. The German Sixth Army breached the Belgian frontier defences, joining rapidly with the airborne forces holding the bridges over the Albert Canal at Vroenhoven and Veldwezelt. In two days the Belgians lost both their main fortified lines, and were left holding a third line of poor fortifications running from Antwerp to Namur.

The ferocity of the German assault in the north confirmed the Allied belief that this was the main German attack, and that evening French 1st Army Group (including the BEF) commenced moving forward to the Breda-Dyle line. Despite reaching southern Holland, French Seventh Army found itself subject to heavy air attack, and lacking armoured support it retreated in the face of 9th Panzer Division. Further south French forces blunted Army Group B's thrust into central Belgium. For two days the French Cavalry Corps under General Robert Prioux, consisting mainly of two Light Mechanized Divisions, delayed XVI Panzer Corps, while French First Army, the BEF, and Belgian forces prepared to defend the Antwerp–Namur line. On 15 May, German Sixth Army failed to breach the Allied line in Belgium, and for a few hours the Allied armies believed that they had blocked the German offensive. The rude awakening came that evening: a disaster had befallen French forces to the south and orders were to retreat.

While Army Group B was invading the Low Countries, Army Group A was moving through the Ardennes. Two Belgian divisions in the area retreated to the northeast, providing no serious opposition, and a screen of French light cavalry divisions was rapidly swept aside. In its movement through the Ardennes, Army Group A suffered more delay from traffic jams than from any Allied military action. By the evening of 12 May, German armour had reached the line of the River Meuse. The Allied High Command still thought that the main German attack was further north, and anyway expected that it would be about three days before the enemy was ready to attempt a river crossing. They were wrong: on the next day the Germans assaulted the last formidable natural defensive line between them and the coast.

In its 75-mile stretch from Sedan to Namur, the River Meuse was wide enough to constitute a formidable natural barrier, and although in the north the river was overlooked by hills on the eastern bank, from Givet to Sedan wooded hills on the western bank dominated the river. However, the six French divisions holding this section of the river were dangerously overstretched, and were mainly third-rate formations with obsolete equipment. In the north, where the river line was less defens-ible, French forces were disorganized, and exhausted, by their move forward into Belgian fortifications which were found to be largely nonexistent. In contrast, the Panzer divisions constituted the cutting edge of the *Wehrmacht*, and their all-arms mechanized structure ensured that they could attempt an immediate crossing, with the Luftwaffe providing the additional firepower necessary to suppress French defences.

Typical of the German assaults over the Meuse on 13 May, all of which were successful, was that of General Heinz Guderian's XIX Panzer Corps at Sedan. This crossing was especially important, as it constituted the southern flank of any subsequent advance. Opposing Guderian was the French 55th Division, a 'Category B Division' of third-rate reservists. Its 140 artillery pieces (over double the normal div-isional complement) could have inflicted considerable losses on the Germans. However, French artillery was neutralized by German airpower, as two Luftwaffe Flying Corps (roughly 1,100 aircraft) delivered a five-hour bombardment. Matters were made worse by French orders, issued in the expectation of a long attritional slogging match, restricting the supply of shells. German infantry crossed the river closely supported by Junkers Ju-87 'Stuka' dive bombers, and by nightfall they had deprived the French of the crucial heights dominating the river. German engineers worked at great speed, building a bridge by midnight; getting XIX Panzer Corps' heavy equipment – especially the tanks – over the river was the key to consolidating and exploiting success. French hopes of preventing disaster rested on launching a rapid counterattack, but both of the French attempts to destroy the German bridgehead were belatedly launched and ineptly conducted. The French 3rd Armoured Division was ready for action just after midday on 14 May, but the attack was postponed and the armour dispersed into a defensive line; a fleeting window of opportunity disappeared, and, unable to reunite, the division fought in ineffective dribs and drabs. By 15 May, XIX Panzer Corps, like the two Panzer corps to the north, had consolidated its bridgehead and prepared to advance to the Channel coast.

The Allied Collapse

On the map the German advance to the coast looked eminently preventable. On both sides of the 'Panzer Corridor', the Allies had substantial forces and only a few German motorized formations defended the Panzer divisions' lines of communication. However, the shock of the German surprise advance made it difficult for the Allies to make a coherent military response. Panzer divisions by-passed, or simply drove through, the courageous but ineffective attempts by Allied forces to halt their progress, reaching the English Channel coast near Abbeville on 20 May. The Luftwaffe dominated the air, launching attacks against Allied forces, hindering military movement, and compounding the chaos on the ground. In the north, Army Group B continued to press Allied forces, and the fall of Antwerp on 18 May undermined the efforts of Allied forces to conduct an organized fighting retreat. Consequently, Allied attempts to attack the Panzer Corridor were spasmodic and poorly coordinated. Although even small successes, such as the British attack against the 7th Panzer Division near Arras on 21 May, caused the Germans great anxiety, bolder councils held sway as they continued to exploit their advantage.

Military disaster quickly exposed Allied differences, as an apparent paralysis set in within their higher commands. On 15 May Britain refused to transfer fighters from home defence to France and King Leopold of Belgium ordered his forces to retreat north rather than west, a clear sign that he was contemplating surrender. Once it became likely that French 1st Army Group would be cut off, the French planned to maintain their forces by sea in the hope of attacking the Panzer Corridor, but the British planned evacuation through the Channel ports, eventually deciding on Dunkirk. On 23 May the BEF abandoned Arras, and on 25 May French First Army made it clear that its losses were so high that no attack could be contemplated. Threatened from all sides, French 1st Army Group held a strip on the Belgian coast and a narrow finger stretching roughly 80 miles down the French–Belgian frontier.

Allied defences were crumbling, on 24 May German Sixth Army broke through the Belgian sector on the line of the River Lys, and it seemed that Dunkirk could only be held a few more days. On 26 May the well-organized British began evacuating Allied forces from Dunkirk, joined rather haphazardly by the French, but on the evening of 27 May the Belgians surrendered, leaving a long gap in the Allied line. The Allied armies began to disintegrate. That the Allies were able to hold on until 4 June, long enough to evacuate over 370,000 soldiers (60% of them British), was due to Hitler halting the overextended Panzers for two days (24–26 May), and to French First Army's defence of the Lille area tying up seven German divisions, three of them Panzer divisions. What seemed the 'miracle' of Dunkirk must not be allowed to obscure the scale of the disaster: roughly 61 Allied divisions had been destroyed, including the best Allied armoured and motorized formations.

On 5 June the German offensive resumed, turning southwards. The French held a new line running along the rivers Somme and Aisne, but the skies were dominated by the Luftwaffe, on the ground the Germans had a numerical edge of over two to one, and their advantage in armour was now in excess of five to one. Too late the French used an effective in-depth defence, showing their previous debacle was by no means inevitable, but they could not prevent the German advance. By 12 June German forces were over the River Seine on either side of Paris, and moving south of Rheims. Completing the defeat, Army Group C with Guderian's Panzers in support, penetrated the barely defended Maginot line in Lorraine. Political crisis and military defeat went hand in hand. The French government collapsed, and on 22 June a new government under the 84-year-old Marshal Philippe Pétain signed an armistice with Germany, acknowledging total defeat.

SEAN MCKNIGHT

Further Reading

Bond, B., *Britain, France and Belgium* (London, 1990)
Cohen, E.A. and Gooch, J. (eds), *Military Misfortunes* (New York, 1990)
Doughty, R., *Breaking Point* (New York, 1990)
Horne, A., *To Lose A Battle* (London, 1979)

the Britain's [illegible] advanced, leaving a long [illegible] line. The Allied [illegible] that the Allies were able only [illegible] [illegible] 27 [illegible] of the [illegible] [illegible] Panzer [illegible] 26 [illegible] of the [illegible] [illegible] [illegible]. What [illegible] of Caen [illegible] including [illegible] Allied [illegible].

[illegible] the Germans [illegible] [illegible]. The [illegible] that [illegible] the Germans [illegible] [illegible] [illegible] [illegible] [illegible] [illegible] Germans [illegible] [illegible] south of [illegible] [illegible] Army Group B [illegible] [illegible] [illegible] [illegible] Marshal Philippe Pétain [illegible] [illegible] Germany [illegible].

Further Reading

Misunderstanding

'No battle plan survives contact with the enemy.'

Field Marshal Helmuth von Moltke

It does not matter how good a battle plan is if it cannot be carried out properly. Many of the classic disasters of military history, and even some of the great 'what ifs', have been the product of potentially good and sensible plans that have gone wrong in the execution. Sometimes this is the fault of the commander who made the plan for not realizing the limitations of his army or his subordinates, and that he might be asking too much of them. Often it is the product of a number of small but tragic errors, which accumulate to produce a disaster.

Waterloo

18 June 1815

The Limits of Enterprise

'I tell you Wellington is a bad general, the English are bad troops and this will be a picnic.' Napoleon

No one has ever doubted the military genius of Napoleon, despite the fact that he came close to defeat in several of his battles. But it takes nothing away from his opponents at Waterloo to say that one of the major puzzles of military history is how he finally came to lose a battle that he should have won, having given himself every chance of doing so, and placing his enemies in a desperate situation. The answer does not lie in any one factor, but in the accumulation of a number of small misunderstandings and errors, starting with the degree of improvisation needed to fight the campaign at all. Against generals and armies of the highest quality, there was a limit to what even Napoleon could do.

The Campaign of 1815

Napoleon's return to France in April 1815 from exile on the island of Elba was one of the greatest gambles in history. Initially it paid off with the flight of the Bourbon monarch Louis XVIII, and Napoleon's return as Emperor of the French. However, the Allies who had defeated Napoleon in 1814 were meeting in Vienna to decide the future of Europe, and all agreed that 'The Ogre', as they called him, must be dealt with once and for all. Within months, Napoleon knew that the Austrian and Russian armies would be approaching the French frontier. But in the meantime, the Anglo-Dutch Army under the Duke of Wellington and the Prussian

Army under Prince Blücher stationed in Belgium were the main threat. Napoleon decided to mount a rapid campaign of attack, aiming to keep these two armies separated and deal with each in detail, a technique that he had used on many occasions before. The early defeats of these Allied armies combined with the capture of Brussels would give Napoleon a great political and military victory, which might well secure his position on the throne of France.

However, although Napoleon's planning of the campaign was masterly, the execution fell below his expectations. On 16 June, Marshal Michel Ney, Prince of the Moskowa, commanding Napoleon's left wing, was halted by Wellington's hurriedly concentrated force at Quatre Bras, south of Brussels. While Napoleon with the right wing of his army gave Blücher's Prussians a severe mauling at nearby Ligny, the Prussians were not destroyed, and were able to retreat to the north towards Wavre and a junction with Wellington's forces. Napoleon fell into an uncharacteristic lethargy on the morning of 17 June, which allowed Wellington to withdraw from his now perilous position at Quatre Bras to the position he had selected for defence: the ridge at Mont-St Jean, to the south of the village of Waterloo. Napoleon's lethargy also delayed the pursuit of the Prussians, to which he had entrusted a third of his army under Marshal the Marquis de Grouchy, which meant that he failed to identify the Prussian routes and intentions. As the French Army marched towards Mont-St Jean and battle with Wellington, Napoleon was convinced that the Prussians were too badly beaten to interfere with his destruction of Wellington next day. Already, misunderstandings between Napoleon and Ney, and Napoleon and Grouchy, had begun to affect his chances of victory.

Napoleon's Plan for Waterloo

Napoleon's plan certainly was not subtle, and did not contain any of the finesse which some of his earlier battles displayed. Much of this was due to Napoleon's contempt for Wellington and his coalition army, despite the warnings of some of his experienced commanders who had encountered Wellington in Spain. Napoleon planned to beat Wellington quickly and decisively by punching through the Anglo-Dutch Army and throwing it off the ridge in a massive frontal attack. He aimed to hold Wellington's attention at the Château of Hougoumont with a secondary attack by

General Count de Reille's Corps, which might even draw some Allied troops away from the centre. At the same time, he planned to use a 'Grande Batterie' of over 80 of his 'belle filles' – his beautiful daughters, as he called his heavy 12-pounder guns – to batter the Allied centre. Once the artillery had softened up the Allied positions, General Count d'Erlon's Corps of 20,000 infantry would attack Wellington's centre, and capture the vital village of Mont-St Jean which lay behind it. With this in French hands, Napoleon could release his cavalry and his main reserve, the Imperial Guard, to destroy Wellington's army.

Wellington's Plan

Wellington's plan was also simple – to stand on the ridge doggedly until the Prussians were able to come to his assistance. Without Prussian help, the day would be lost. However, Wellington deployed his troops skilfully to maximize their effectiveness. The various Allied contingents of Brunswickers, Hanoverians, Nassuers, and Dutch-Belgians were corseted by British units placed on either side, and all were deployed to make the best possible use of the ground. On Wellington's right, the Château of Hougoumont, garrisoned by companies of his Foot Guards, provided a strong bastion against any French movement around his right flank. To guard against such a movement, the bulk of Wellington's reserves were placed on his right at the outset of the battle. To the west, there were further Dutch-Belgian troops two miles away at Braine L'Alleud, while 17 miles away there were 17,000 troops at Hal. These troops, while they were not engaged in the battle, kept open Wellington's line of retreat to the English Channel ports, and they provided a long stop against any wide French turning movement to his right. The farm of La Haye Sainte, occupied by Major Baring's battalion of the King's German Legion (troops of German nationality in British service), lay in Wellington's centre, and although there was neither the time nor the tools to properly fortify the farm complex, La Haye Sainte provided a key position. Always careful to minimize casualties, Wellington stationed his main line *behind* the ridge which ran from La Haye Sainte to the villages of Papelotte, La Haie, and Frischermont, which guarded his left flank. These villages were surrounded by rough ground, including woods and a stream, which made the going difficult. Wellington's position and his deployment thus formed the major part of his planning. He had insured himself against any French turning movement, while placing his army on a defensible position which constricted any obvious lines of attack to two – either between Hougoumont and La Haye Sainte, or between La Haye Sainte and Papelotte.

In order to fight at all, however, Wellington needed assurances that the Prussian Army would march to support him. Perhaps the key decision was made by Blücher, who gave his word to Wellington that he would march to his aid. But while Blücher gave his whole-hearted support to Wellington, his chief of staff, General Graf N von Gneisenau, was suspicious of Wellington and his motives. Gneisenau was determined not to make a forced march to Mont-St Jean, only to find that Wellington had retreated, leaving the Prussians exposed before Napoleon's full forces. For this reason, he scheduled General F W von Bulow's IV Corps to lead the march, even though his corps was the furthest away from Wellington. This did delay the Prussian arrival, but ensured that, on arrival, the most powerful corps went into action first.

The Start of the Battle

The Belgian weather had not been kind over the previous few days, and on the night of 17–18 June there had been torrential rain. This meant that the heavy Belgian soil was very muddy, and hindered the movement of Napoleon's artillery, which was vital for the success of his plan. It was mainly this mud, which was axle deep in places, which persuaded Napoleon to delay his attack until the ground had dried sufficiently to allow his artillery to deploy effectively. However, while the mud was a major factor in Napoleon's decision, it was also true that many French units were still marching to their forming up places, and men were being gathered in from their night-time foraging and bivouacs. In any event, Napoleon could not have begun the battle much before he did, at 11.00 a.m., but the loss of these morning hours was to prove crucial to the outcome of the battle.

The battle began with Reille's diversionary attack on Hougoumont. Although orders specifically ruled out an attack on the formidable château, Prince Jerome Bonaparte, Napoleon's brother, became carried away and committed more and more troops of his division to its capture. Ultimately, the fight for Hougoumont became a battle within the battle, and the bitter conflict raged all day. Despite all their efforts, the château was never captured by the French. This was the first of the misunderstandings on the battlefield which beset Napoleon.

D'Erlon's Attack

While the battle raged at Hougoumont, the Grande Batterie had pounded the Allied line, and at 1.00 p.m. this preparation was deemed sufficient. The 20,000 veteran troops of d'Erlon's Corps, supported by one brigade of cuirassiers (heavy cavalry with breastplates) under General Count Milhaud, advanced to the attack. Unfortunately, d'Erlon, having consulted his divisional commanders, ordered the troops to use an unwieldy and archaic formation. This was the *Colonne de Division*, in which each battalion in the division was deployed in line, one behind the other. While this maximized the firepower of the lead battalion (as had been the intention – d'Erlon had suffered at the hands of British lines in the Peninsula), it made it nearly impossible for the battalions to form square to defend against cavalry if necessary. This deployment was to prove disastrous. Only one division under General Baron Durutte, tasked with clearing Papelotte and La Haie, was deployed in the hand-ier, and more conventional, battalion columns.

As the other three divisions of d'Erlon's Corps pushed up the slope to the Allied positions, they came under heavy artillery fire which ploughed large gaps in their ranks, but the French formations did not falter. One division surged around La Haye Sainte, while the other two divisions crested the ridge and began a fire-fight with the Peninsula veterans of the British 3rd Division under Lieutenant General Sir Thomas Picton. The formidable French columns, deployed to maximize their firepower, began to win the fight. This was the first crisis of the day for Wellington, for Picton's men were the only Allied troops between d'Erlon and Brussels. As Picton fell leading his reserves into the battle, shot through the head, the day hung in the balance.

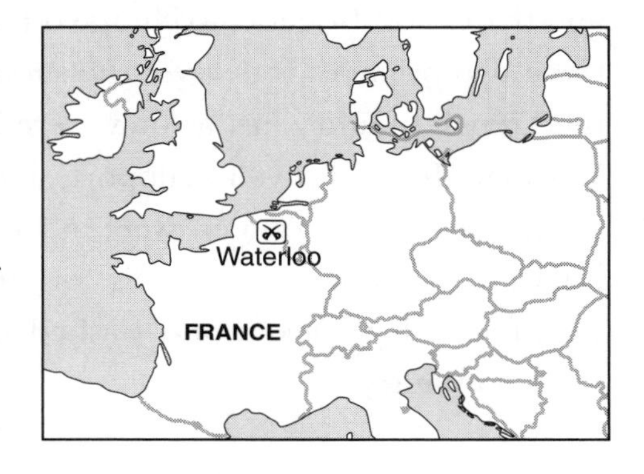

It was General the Earl of Uxbridge, Wellington's second in command and cavalry commander, who seized the moment by

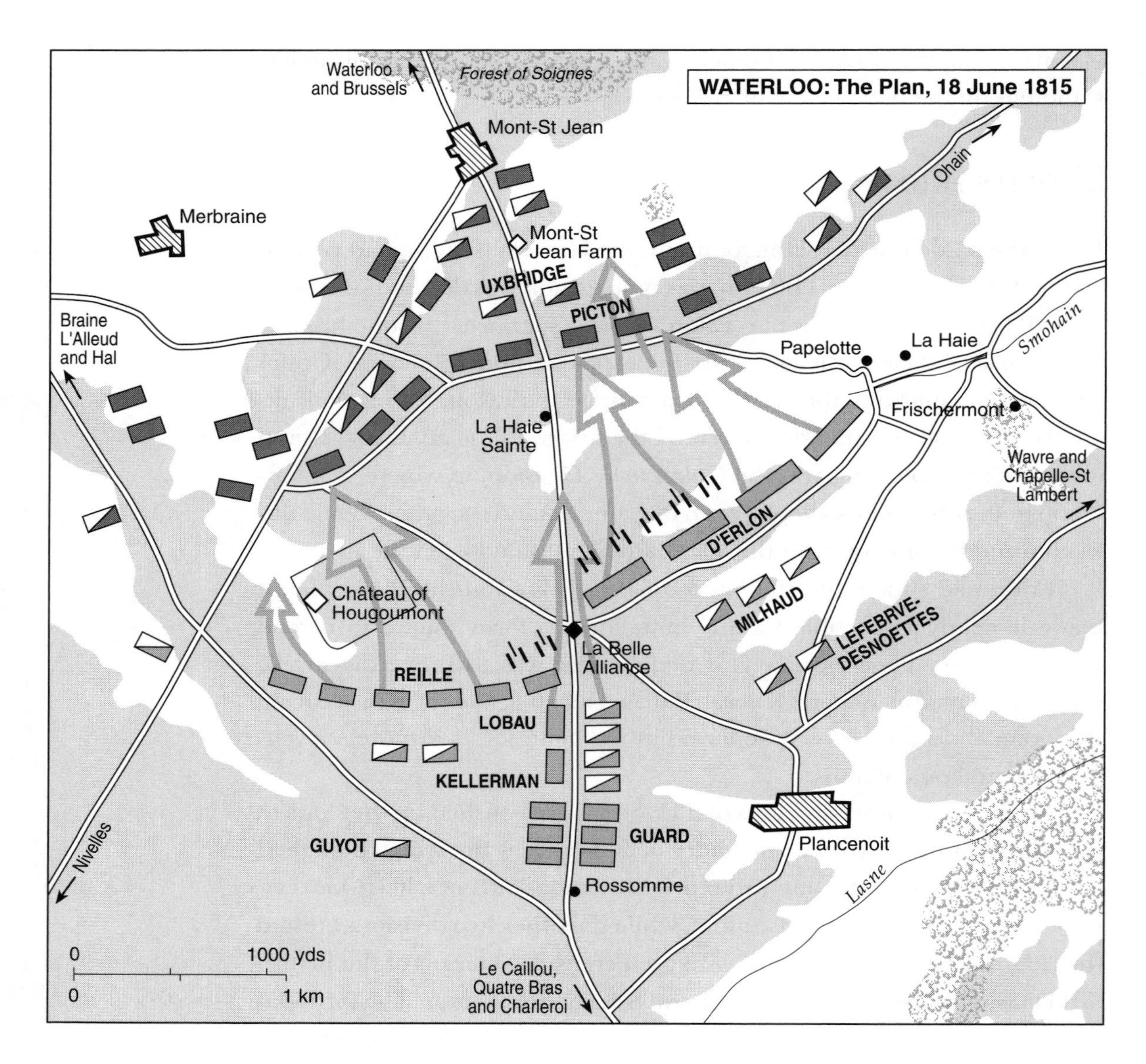

ordering his two British heavy cavalry brigades, the Household Brigade and the Union Brigade, to charge d'Erlon's infantry. With perfect timing, these two brigades rode over the supporting French cuirassiers and into the French infantry just as they were beginning to gain the upper hand. Without effective cavalry support, and unable to form square to protect themselves, d'Erlon's men were at the mercy of the British cavalrymen. Literally in minutes, d'Erlon's formations dissolved into chaos as the British cavalry hacked and slashed their way through the defenceless French infantry.

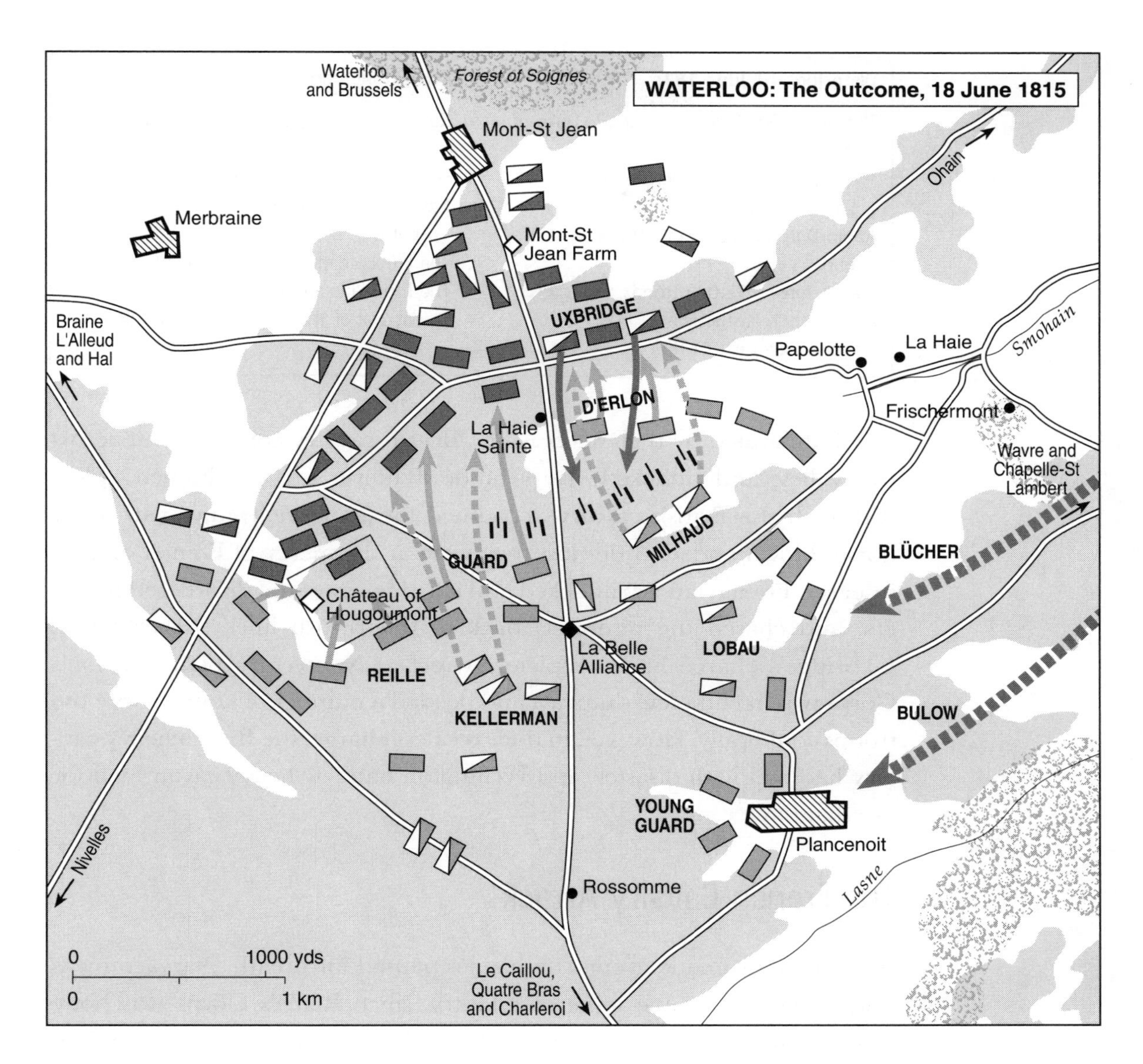
WATERLOO: The Outcome, 18 June 1815
Waterloo and Brussels
Forest of Soignes
Mont-St Jean
Merbraine
Mont-St Jean Farm
Ohain
Braine L'Alleud and Hal
UXBRIDGE
Papelotte
La Haie
Smohain
D'ERLON
La Haie Sainte
Frischermont
Wavre and Chapelle-St Lambert
MILHAUD
GUARD
BLÜCHER
Château of Hougoumont
La Belle Alliance
LOBAU
REILLE
KELLERMAN
BULOW
YOUNG GUARD
Plancenoit
Lasne
Nivelles
Rossomme
0
1000 yds
0
1 km
Le Caillou, Quatre Bras and Charleroi

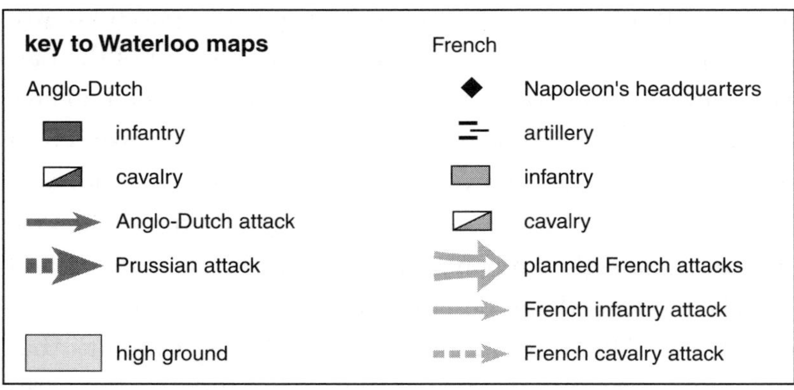
key to Waterloo maps
Anglo-Dutch
infantry
cavalry
Anglo-Dutch attack
Prussian attack
high ground
French
Napoleon's headquarters
artillery
infantry
cavalry
planned French attacks
French infantry attack
French cavalry attack

Waterloo, 18 June 1815	Critical Moments
The Anglo-Dutch Army under the Duke of Wellington	Reille's failed attack on the Château of Hougoumont fails
The French Armée du Nord under the Emperor Napoleon Bonaparte	D'Erlon's attack in the centre is repulsed
Anglo-Dutch forces: 68,000 troops, 156 guns	Ney's cavalry charges against the Allied centre
French forces: 72,000 troops, 246 guns	The French capture La Haye Sainte farm
Anglo-Dutch casualties: 22,000	The Prussians arrive from the east
French casualties: 41,000	The attack of the Imperial Guard fails

Exhilarated by their success, the British cavalry continued on across the valley and attacked the Grande Batterie on the French ridge. Although the British cavalry accounted for 20 French guns, with their horses blown and all order lost they were at the mercy of French cavalry reserves. French lancers attacked and killed many of the disordered troopers, and chased the remnants back to the British lines. Nevertheless, Uxbridge's charge had completely wrecked Napoleon's plan: d'Erlon's Corps was in complete disarray, and needed a number of hours before the troops could rally. However, in their reckless charge, the British heavy cavalry had met with disaster, and Wellington had few heavy cavalry left for use that day.

The French Cavalry Attack

After these dramatic events, both sides paused for breath. Napoleon was now without a usable force of infantry, since Reille's Corps was hotly engaged around Hougoumont, while d'Erlon's Corps was still reforming. The only infantry left available to him were his superb Imperial Guard, and he was not about to use them before the decisive moment of the battle had arrived. Ney managed to scrape a few battalions to attack La Haye Sainte and the ridge line, but these were easily repulsed. However, given the increasing casualties caused by the constant French cannonade, Wellington ordered that his army should fall back a hundred paces to give it better protection against the French guns. Ney saw this movement, and the stream of Allied wounded and deserters moving towards the Forest of Soignes behind the Allied position, and mistakenly thought that Wellington might be retreating. In order to take advantage of any withdrawal he ordered a brigade of cuirassiers to probe the Allied line. Unfortunately, the brigade commander's immediate super-ior queried the order, and Ney angrily ordered the whole of Milhaud's cuirassier corps to attack. Milhaud, dismayed at such orders, asked the commander of the Imperial Guard Light Cavalry, General Charles Lefebrve-Desnoettes, to support him. Soon some 4,000 French cavalry were advancing across the valley to attack the Allied line.

There was great surprise up and down Wellington's line that Napoleon was about to launch an unsupported cavalry attack against firm infantry, although such an attack *had* worked for Napoleon at Eylau in 1807, and at Borodino in 1812. Nonetheless, the Allied infantry were undaunted and closed up into a chequerboard pattern of 20 squares (many containing more than one regiment or battalion) to receive the cavalry charge. As the French cavalry advanced, magnificent in their steel breastplates and the horsehair manes on their helmets, they came under the concentrated fire of the Allied gun batteries. Great gaps were torn in their ranks, only to be filled again. Eventually the Allied gunners, having caused serious casualties, had to run to the protection of the infantry squares. However, once the French cavalry reached the squares, there was little that they could do. Napoleon had detached all of the horse artillery batteries from his cavalry, which meant there were no French guns which could unlimber at short range and blast gaps in the squares with canister shot, nor were there any French gunners who could spike the Allied guns. The French cavalry constantly checked for weaknesses in the Allied squares, firing their carbines and pistols at short range. But the Allied infantry, using controlled volleys, were able to empty many saddles. Raging at their impotence, Milhaud's brave men eventually retreated, pursued by the remnants of the Household and Union Brigades.

After learning of Ney's orders, Napoleon had snapped, 'He has compromised us as he did at Jena.' But far from countermanding Ney's orders, Napoleon compounded them by ordering both General Count Kellerman's cuirassier corps and General Baron Guyot's Heavy Cavalry of the Guard into the attack. Virtually all of Napoleon's magnificent cavalry was now committed into unsupported attacks against the Allied line, with predictable results. The French cavalry attacked again and again – at least five separate charges were made but with little gain, although they were able to hack down a couple of Allied squares. In fact during these hours, the Allied infantry began to breathe a sigh of relief when the French cavalry attacked, because it gave them respite from the terrible bombardment of the French guns. Packed into dense, immobile square formations, the infantry made excellent targets for the French gunners. Casualties began to mount alarmingly, and more and more battalions had to combine to form a single square.

Eventually, Ney realized that some of Reille's Corps were not engaged, and he led forward a division against the Allied line between Hougoumont and La Haye Sainte, covered by the remnants of the cuirassier regiments. Unfortunately, this part of the line had been recently reinforced, and the division advanced into a storm of converging artillery and musket fire. With the failure of this attack, the French redoubled their cannonade, and sent forward as many infantry skirmishers as they could to harass the Allied line. This fusillade caused many casualties, and many Allied units began to lose heart.

The Capture of La Haye Sainte

It was only after it became clear that the cavalry attacks had failed that Napoleon ordered Ney to take La Haye Sainte at all costs. Finally, Napoleon had realized that this farm complex was the key to the battle. Ney scraped together a few thousand troops from d'Erlon's Corps and supported them with elements of cuirassier regiments. The garrison, who had fought doggedly for most of the day, were not only tired but almost out of ammunition, their re-supply having failed to reach them. Unfortunately, two battalions of the King's German Legion sent to reinforce the position advanced in line, and were cut to pieces by the French cuirassiers. After fighting to the last round, Baring was forced to evacuate the farmhouse and leave it in French hands.

With the fall of La Haye Sainte at 6.00 p.m., there was now a gap in Wellington's centre. Victory was within Ney's grasp, but he needed more troops to exploit this into a breakthrough. Ney hurriedly dispatched a messenger to Napoleon to ask for more infantry. But Napoleon simply retorted, 'Troops? Where do you expect me to find them? Do you think I can make some?' The fall of La Haye Sainte was the critical moment of the battle; to win the day Napoleon needed to release the 14 battalions of his Imperial Guard in an all-out attempt to destroy Wellington's army.

Instead, Napoleon's attention was drawn away to his right where the Prussians were now pushing forwards against Plancenoit.

Since 3.00 p.m., on Napoleon's right the small French corps under General Count Lobau had been holding out against increasing Prussian pressure, but had been pushed out of Plancenoit by Bulow's advancing men by 6.00 p.m. This was extremely dangerous for Napoleon. The Prussian line of advance threatened to cut off his retreat, and with Prussian cannon balls now bounding across the Brussels–Charleroi road, the Emperor decided to deal with the threat. He sent in the Young Guard, supported by two battalions of the Old Guard, who retook Plancenoit at the bayonet without firing a shot. With the situation thus stabilized, Napoleon was able to turn his attention to Wellington again, but by now it was nearly 7.00 p.m. and the chance to beat Wellington had slipped away. Wellington had stripped his flanks to guard his centre, and the gap which had been so dangerous had been sealed.

The Attack of the Imperial Guard

Napoleon still believed he could snatch victory by launching his Guard against the tired and much reduced Allied army which still lined its ridge. He led 12 battalions of his finest troops, with the Middle Guard in front and the Old Guard in the second line, up to the attack. To exhort his own tired troops into one last effort, he sent staff officers along the line who shouted, 'Vive l'Empereur! Soldats! Voila Grouchy!' Even though the dark masses which could be seen over on the right were advancing Prussians and not Grouchy's expected forces, the psychologic-al trick worked, and French soldiers all along the line roused themselves in a final supreme effort.

Unfortunately, in one last French misjudgement, the Guard attack was sent forwards between Hougoumont and La Haye Sainte, rather than beside La Haye Sainte where it would have gained support. Instead, the Guard marched up the slope into a hail of Allied artillery fire, and where the best and freshest troops were waiting, a British Brigade of Foot Guards under Colonel Maitland and a Netherlands Brigade under Colonel Detmer. Napoleon was persuaded at the bottom of the slope not to lead his final attack personally, and the Guard marched away from him into the thick smoke which hung over the ridge. The Guard marched with perfect precision up the slope, but found the Allied defenders undaunted. Volley after volley crashed into the French formations, and incredibly, the Imperial Guard, never before defeated in battle, began to retire.

The Collapse of the French Army

The cry 'La Garde Recule!' was too much for the French, now only too aware of the Prussian advance which had broken through the hinge between d'Erlon's and Lobau's men near Papelotte. Exhausted after a day of bloody fighting, and now attacked powerfully from an unexpected direction, Napoleon's army dissolved into panic. Wellington, with a wave of his hat, ordered his army into a general advance against the beaten French. Blücher also sent his hussars and Uhlans (Prussian lancers) into a merciless pursuit.

As the once proud Armée du Nord dispersed into a mass of fugitives, only the Imperial Guard managed to form square and slowly march off the field, protecting their Emperor as they did so. However, the pressure of the Allied advance was such that Napoleon had to take to his carriage and flee for his life. Even the Guard was finally overtaken and forced eventually to surrender. Wellington and Blücher met near the inn called La Belle Alliance, only recently Napoleon's headquarters, and agreed that the fresher Prussians should pursue the beaten French. One of the most decisive battles in history was over. Wellington was later to say that the only sadder thing than a battle won was a battle lost.

NIALL BARR

Further Reading

Barnett, C., *Bonaparte* (New York, 1978)
Chandler, D., *The Campaigns of Napoleon* (London, 1966)
Longford, E., *Wellington* (London, 1969)

The Battle of Balaklava

25 October 1854

The Misunderstood Order

'It is difficult, if not impossible, to do justice to the feat of these mad cavalry, for, having lost a quarter of their numbers and being apparently impervious to new dangers and further losses, they quickly reformed their squadrons to return over the same ground littered with their dead and dying. With such desperate courage these valiant lunatics set off again, and not one of the living – even the wounded – surrendered.'

Lieutenant Kozhukhov, Russian Artillery

In the heat of battle a cool head is the first requirement of any general. The Charge of the Light Brigade at Balaklava, in what the British called the 'Russian War' (1854–56), the most famous 'military blunder' in history, resulted from the lethal combination of an imprecise order, a hot-headed messenger, and an angry general. Unable to see the full picture of the battle, and unable to think clearly, the British commander Lord Lucan ordered a cavalry brigade to attack Russian field guns and cavalry at the end of a valley which was commanded from both sides by guns and infantry. Despite the odds this astonishing force broke through the gun line, drove the enemy cavalry off the field in panic, and retired in reasonable order back to their starting point.

The Crimean Campaign

Britain and France declared war against Russia in March 1854, to preserve the Ottoman Turkish Empire from Russian aggression. In September their combined armies, 50,000 strong, had landed in the northern Crimea. They planned to capture and destroy the Russian naval base at Sevastopol in a rapid, combined operation, and leave the peninsula before the winter. Everything depended on speed, inter-Allied and inter-service cooperation, and a compliant enemy. The Russian theatre commander, Prince Menshikov, gave battle in a strong defensive position at the River Alma on 20 September, but was heavily defeated. Rather than shut his army up in Sevastopol he withdrew into the inter-ior, leaving a small military garrison and the sailors of the fleet to hold the city. He planned to build up his army, and then move onto the flank of the Allied operation against Sevastopol, diverting them from their object.

Meanwhile, the Allies had moved around to attack the southern sector of Sevastopol. On 17 October they finally opened a heavy bombardment. The British quickly overpowered the Russian defences in their sector, but the bombardment stalled when the main French magazine exploded. Well aware that the Allies would soon resume their attack, and under pressure from Tsar Nicholas I to regain the initiative, Menshikov committed his field army before it had received sufficient reinforcements to achieve a decisive success. He planned a limited, probing attack aimed at the British supply base and harbour at Balaklava. This would achieve the maximum diversion with the minimum effort.

Menshikov, reinforced by Lieutenant General Prince P P Liprandi's 12th Infantry Division, which had come from Bessarabia by forced marches, with elements of the 17th Division and additional cavalry, now had 65,000 men. He could expect another 25,000 shortly, giving him a significant advantage over the Allies, who had been reinforced to 75,000 men. The Allies were jointly commanded by the British General Lord Raglan and the French Marshal F de C Canrobert.

The Battlefield

The main feature of the Balaklava battlefield was the high ground of the Sapun Escarpment (or Crimean Uplands) which dropped away quite steeply onto the Plain of Balaklava. This was divided into two valleys by the Causeway Heights, along which ran the Woronzoff road. To reach Balaklava the Russians would have to cross the River Tchernaya at the Traktir Bridge, move across the Fediukhine Heights, and enter the North Valley. The Allied front line was made up of a series of redoubts on the Causeway Heights, which were manned by newly arrived Turkish militia units, with some British guns. Beyond them lay the defences of Balaklava, an outer line held by the 650 men of the British 93rd Highlanders (later known as the Argyll and Sutherland Highlanders), and an inner position manned by about 1,000 Royal Marines and a simi-lar number of Turkish troops, all under the overall command of General Sir Colin Campbell. As they were not required for siege operations, the British Cavalry Division, commanded by the Earl of Lucan, was encamped at the foot of the escarpment to cover the open flank.

The Battle of Balaklava

The Russian plan was for their infantry to capture the redoubts, and then send their cavalry to attack Balaklava. On the way, they expected to engage the British cavalry. Menshikov, who was a sound strategist but a fatalistic and inflexible battlefield commander, left the tactical command to energetic and popular Prince Liprandi. At dawn on 25 October the Russians, already across the Traktir Bridge, had their artillery in position to attack the redoubts. Within 90 minutes, the first redoubt had been stormed, and most of the defenders bayoneted. Thirty minutes later three more redoubts had been taken, most of the militia men taking to their heels.

When the firing started Raglan, who was responsible for the eastern flank of the Allied position, ordered his 4th Division and Guards' Division down from the British camp. This required a two-hour march. General Sir George Cathcart, whose 4th Division had been ordered down four days earlier on a false alarm, delayed while his men had breakfast. Cathcart, physically and operationally short-sighted, would be only one of

Raglan's problems that day. While he waited for the infantry, Raglan, who had a commanding view of the plain from the edge of the escarpment, watched in frustration as the Russians began to occupy the redoubts. He was forced to rely on the forces to hand, of which only the cavalry was mobile. His plan was to hold up the Russians without committing the cavalry too far. Like his old chief, the Duke of Wellington, he had no faith in the tactical control of his cavalry commanders, and recognized their propensity to charge at the slightest excuse.

While Raglan was not a great commander, he did have outstanding positional sense. He would invariably find the key point of the battlefield, and knew how to exploit it. At the Battle of the Alma he had ridden right through the Russian centre, and called up guns to exploit the high ground he had occupied. At Balaklava, the cavalry stood at the western end of the South Valley, facing the causeway. The Highlanders, with some Turkish troops, occupied a rise in front of the camp at Kadikoi. The men were held on the reverse slope of the rise, as was customary. When four squadrons of Russian hussars, with Cossack cavalry supports, advanced toward Balaklava they were met by a heavy and accurate fire from both the rifle-armed Highlanders, who suddenly appeared on the crest of the rise about 250 yards ahead of them, and their artillery support. This was the 'Thin Red Streak' celebrated by the war correspond-ent W H Russell (he changed it to 'Thin Red Line' in the book version of his dispatch).

As they passed over the Causeway Heights into the South Valley, the main body of Russian light cavalry, about 2,000 hussars and Cossacks, then encountered the advancing Heavy Brigade of Lucan's Cavalry Division, commanded by General Sir James Scarlett. Although heavily outnumbered, Scarlett ordered 300 of his troopers, from the 2nd Dragoons (known as the Scots Greys) and the 6th (Inniskilling) Dragoons, to charge uphill into the Russians. The Russian cavalry stopped, deployed their line and began to fire. This was a fatal mistake. The British cavalry, big men superbly mounted on large horses, towered over the hussars and drove through them exchanging blows at close quarters. With the advantage of height and weight, the Scots Greys and Inniskillings proved irresistible. The Russians were driven from the field with 270 killed and wounded, some by the flanking British horse artillery fire. The Russians did not have any heavy cavalry in the Crimea, and their newly raised light cavalry and Cossacks were never intended to be a match for the British heavies. That day the Heavy Brigade had only 10 killed and 98 wounded, and most of these were incurred later by the two regiments that supported the Light Brigade. The Russians had kept their most effective force, a regiment of lancers, in reserve. The British pursuit ended when it came under fire from Russian guns. The Light Brigade was perfectly placed to charge the flank of the retiring Russians, but its commander, Lord Cardigan, would not act without orders.

The Charge of the Heavy Brigade, a fine achievement against heavy odds, finished what Campbell's Highlanders had started. It stopped the Russian advance, making up for the slow movement of the 4th Division. The battle might well have petered out at this point. The Russians had achieved a useful diversion, but found the defences of Balaklava too strong, while the British had been forced to withdraw troops from the siege lines.

The Mistake

Raglan was a conventional British general; he wanted to recover the redoubts. Having stopped the half-hearted Russian attack on Balaklava, he ordered Cathcart to attack the three redoubts that the Russians still held. Cathcart refused. When the orders were repeated, Cathcart moved very slowly toward the nearest redoubt. Raglan then ordered Lucan's Cavalry Division to recover the Causeway Heights, if the opportunity arose, with the infantry in support. But Lucan, unable to see the infantry support, waited. About an hour after the Heavy Brigade action, the Russians began to remove the captured British guns from the redoubts. Raglan, well aware that Wellington had never lost a gun, sent the best available horseman, Captain Lewis Edward Nolan, of the 15th Hussars, to ride down the steep slope of the Sapun Heights with an order for Lucan to attack. The order, as written by Raglan's effective chief of staff, General Sir Richard Airey, was not a masterpiece of clarity. It read as follows: 'Lord Raglan wishes the cavalry to advance rapidly to the front – fol-

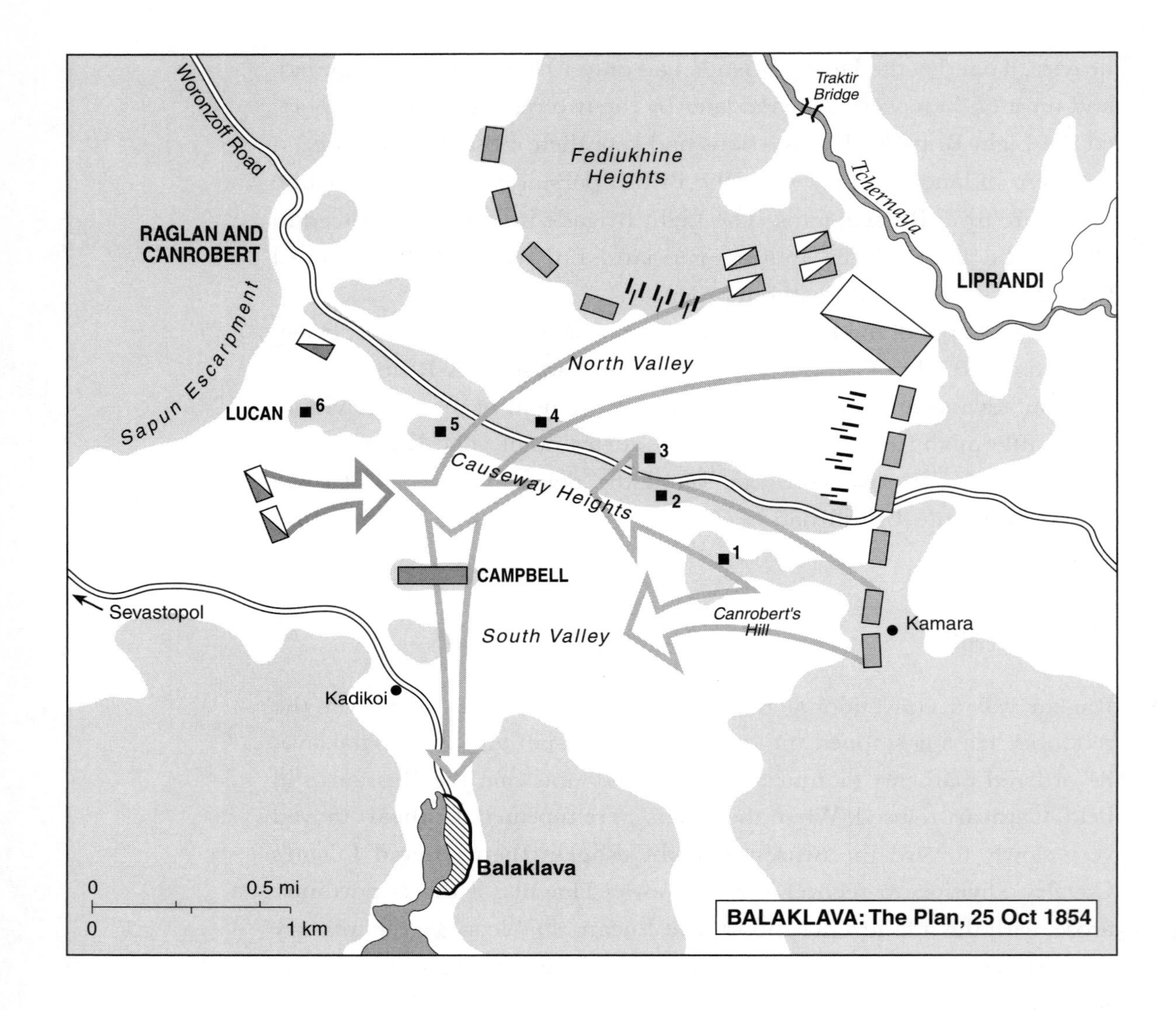

low the enemy and try to prevent the enemy carrying away the guns. Troop Horse Artillery may accompany. French cavalry is on your left. Immediate.' This was supported by an additional, verbal message for Nolan to give to Lucan: 'Tell Lord Lucan the cavalry is to attack immediately.' When Lucan requested further elaboration, being unable to see the positions that were visible to Raglan, who occupied higher ground, Nolan made a violent gesture toward the redoubts and sarcastically observed, 'There, my Lord, is your enemy, there are your guns!' Nolan's contemptuous, dismissive gesture covered the redoubts *and* the main Russian position. Lucan should have known which guns were in question. He had seen

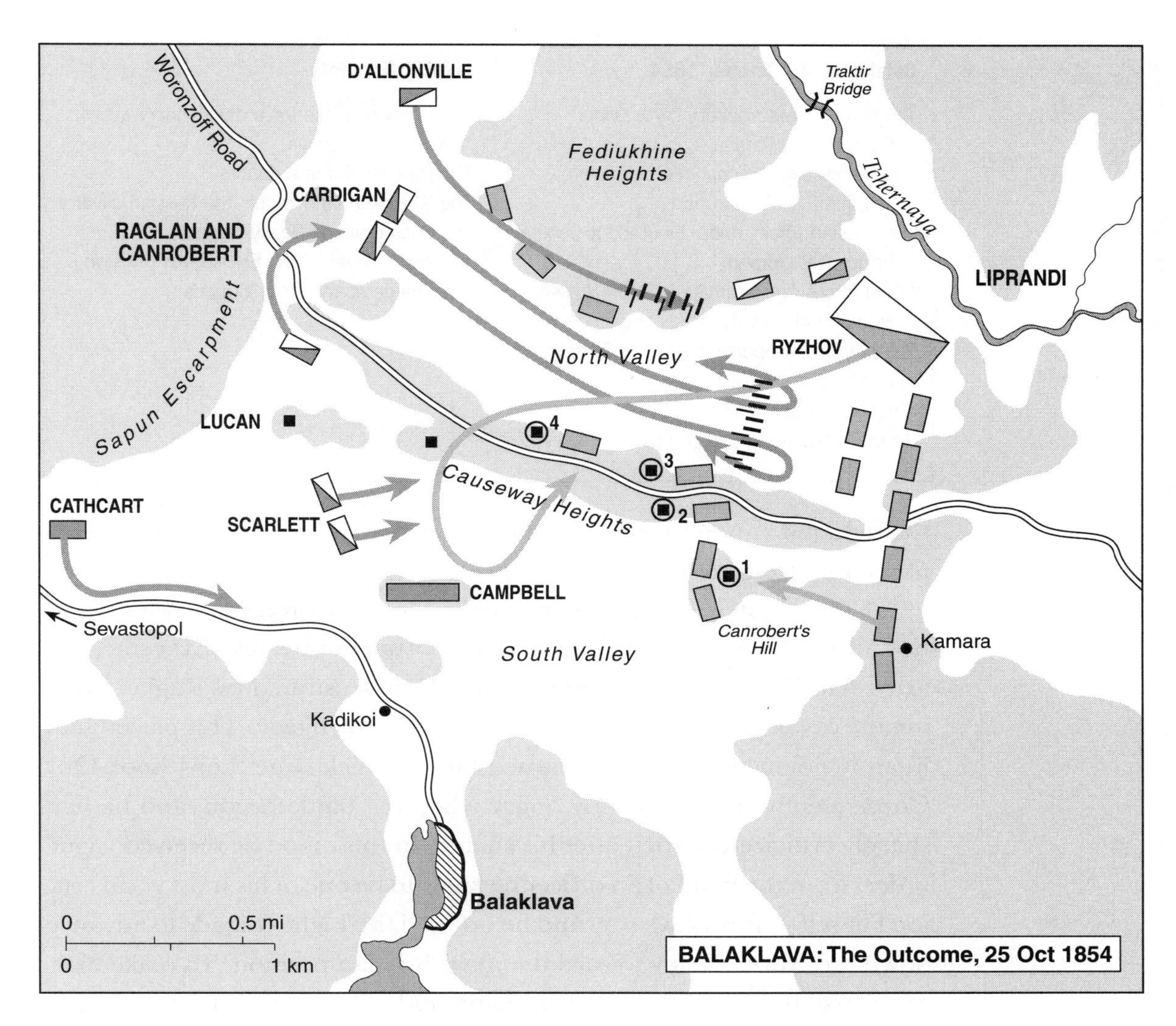
D'ALLONVILLE
Traktir
Bridge
Woronzoff Road
Fediukhine
Heights
Tchernaya
CARDIGAN
RAGLAN AND
CANROBERT
LIPRANDI
Sapun Escarpment
RYZHOV
North Valley
LUCAN
4
3
Causeway Heights
2
CATHCART
SCARLETT
1
CAMPBELL
Sevastopol
Canrobert's
Hill
Kamara
South Valley
Kadikoi
Balaklava
0
0.5 mi
0
1 km
BALAKLAVA: The Outcome, 25 Oct 1854

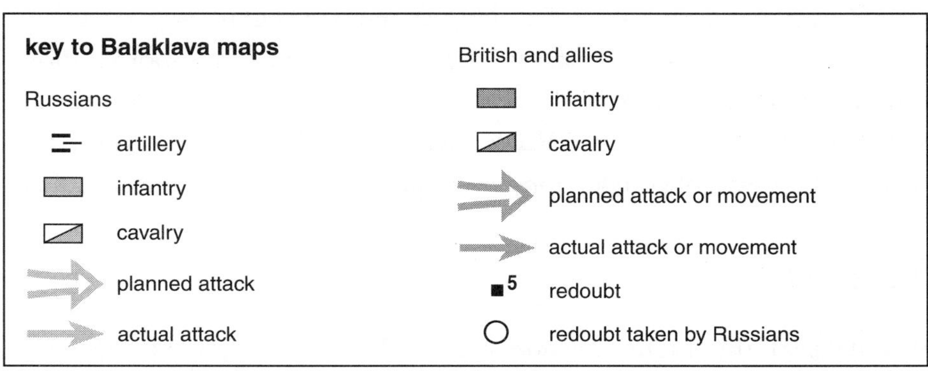
key to Balaklava maps
Russians
artillery
infantry
cavalry
planned attack
actual attack
British and allies
infantry
cavalry
planned attack or movement
actual attack or movement
5
redoubt
redoubt taken by Russians

Balaklava, 25 October 1854	Critical Moments
The British Expeditionary Force under General Lord Raglan	The Russians make for a diversionary attack towards Balaklava
The French Expeditionary Force under Marshal F de C Canrobert	The capture of the redoubts
The Russian forces under Lieutenant General Prince P P Liprandi	The 'Thin Red Line' repels the Russian cavalry
Allied forces: approximately 2,000 British, French and Turkish troops engaged	The Charge of the Heavy Brigade
Russian forces: approximately 3,000 troops engaged	The slow arrival of the British 4th Division
Allied casualties: 658 troops	The Charge of the Light Brigade
Russian casualties: 540 troops	

the redoubts captured earlier in the day, and had previously been ordered to recover the Heights.

Unfortunately Lucan was a volatile, hot-tempered man; what sense he had quickly gave way to rage and bluster under pressure. Ever since the army had landed in the Crimea he had been frustrated by Raglan holding the cavalry back when there were chances to engage. This placed him in an invidious position, and earned him the nickname 'Lord Look-On'. Consequently he was already angry when the battle began, and he had sharply criticized Scarlett after his charge! In this mood he received vague orders from the hand of an officer he knew to be one of his most vocal critics. His self-control gave way and he ordered the Light Brigade to advance down the North Valley toward the main Russian position. To make matters worse the Light Brigade was commanded by Lucan's brother-in-law Lord Cardigan, a man he loathed. When Cardigan, who returned Lucan's hatred with interest and deeply resented his subordination, remonstrated that there were batteries on three sides of the valley, Lucan declared, 'I know it, but Lord Raglan will have it. We have no choice but to obey.' This was, of course, nonsense; but Lucan was too angry to think.

The Charge

At Balaklava the Light Brigade consisted of five regiments: the 8th Hussars and 11th Hussars, the 4th Light Dragoons and 13th Light Dragoons, and the 17th Lancers, totalling 661 men. They were veteran troops, highly trained, mounted on the best horses, and itching for a fight after being left to watch the Heavy Brigade win the glory of the morning. The brigade advanced in three lines, two regiments wide, each regiment in two ranks, with the 11th Hussars forming the second line. The formation was approximately 145 yards wide and 600 yards deep. Cardigan led the formation alone.

Advancing down the North Valley towards the Russian batteries, with the high ground on both flanks held by Russian infantry and artillery, the Light Brigade came under heavy fire from three directions. At this moment Nolan galloped past Cardigan, gesticulating wildly, and was killed by a shell splinter through the heart. It is almost certain that he had realized that the brigade was heading in the wrong direction, and was attempting to redirect it when killed. His efforts proved fruitless. Cardigan continued, completely unmoved. Unlike Lucan he remained perfectly calm throughout the day.

On the Fediukhine Heights to the north of the valley were 14 guns and 8 battalions of Russian infantry; on the Causeway Heights to the south were 32 guns and 11 infantry battalions; while at the end of the valley were the 8 guns of Number 3 Don Cossack Field Battery and 4 regiments of Russian hussars and Cossacks. Under fire the pace of the Light Brigade gradually increased, and cohesion began to slip, not least because the ground sloped downhill. In the third line the 8th Hussars maintained their cohesion, but lost contact with the other regiments. Lucan led two regiments of the Heavy Brigade, the 1st (Royal) Dragoons and the Scots Greys into the North Valley, but recovered enough of his wits to keep them out of the battle. They remained halted, under long-range fire, at the end of the valley throughout the action. They were supported by a brilliant small-scale charge along the Fediukhine Heights by four squadrons of the French 4th Chasseurs D'Afrique, light cavalrymen who were veterans of the Algerian war. Two of their formations, led by General d'Allonville and Major Abd-el-Al, silenced the 14 Russian guns that were firing on the Heavy Brigade at the cost of 10 killed and 28 wounded. They also covered the flank of the Light Brigade as it withdrew.

As the British approached the guns at the head of the valley, the Russian Cavalry commander, General Ryzhov, ordered his four regiments of hussars and Cossacks to cover the guns. Not only was this too late, but the Cossacks in particular had no desire to face the imposing formation bearing down on them. They broke and fled, taking the Ingermanlandsky

Hussars with them. The Light Brigade reached the Cossack guns at the charge, cutting down the few gunners who had not fled. Cardigan rode right through the position, and then retired down the valley. He considered that a general, having led his men into action, had no business risking himself in a mêlée. The remaining men of the first line continued, charging into the retiring mass of Russian cavalry, driving them back some distance. They were supported by the 11th Hussars, but they gradually lost impetus, and were forced to retire. A Russian lancer regiment, which had been waiting between redoubts 2 and 3, attempted to cut them off before they could get back to the gun positions, but were swept aside by the late arrival of the 8th Hussars.

On reaching the position held by the divisional horse artillery and the Heavy Brigade, Cardigan asked after his old regiment, the 11th Hussars, and rode back down the valley to find them. Returning with this unit he was cheered by the Heavy Brigade. The Light Brigade left the action at a slow pace, harried by Cossacks; their horses were completely blown by the charge, most of them were wounded, and many of the men were also injured. One officer, four noncommissioned officers, and one private were awarded the newly instituted Victoria Cross for bravery. It was, under the circumstances, a meagre distribution.

The entire charge had lasted no more than 20 minutes from start to finish. When Cardigan mustered his brigade there were only 195 mounted men. He then thanked the French cavalry and defended his conduct to Raglan. Lucan also refused to take any blame, although he, more than anyone, was responsible. The total casualties were 118 killed and 127 wounded; 45 wounded and unhorsed officers and men were taken prisoner, and 362 horses were killed or subsequently destroyed. Most of the casualties were caused by artillery fire.

Prince Liprandi, who spoke English, interviewed the prisoners. He took some convincing that the entire brigade had not been drunk. They were, in fact, stone cold sober, having been mounted since dawn, and were rather hungry. He was deeply impressed by the stature and composure of the prisoners. The morale effect on the Russians of the discip-line, courage, and resolve of the Light Brigade was immense. For the rest of the war Russian cavalry refused combat with the British, even when vastly superior. Long afterwards the fact that a single, under-strength brigade of light cavalry had captured a battery of guns and driven off a larger body of Russian horse was the admiration of Europe. As one observer, the French General Bosquet, declared, *'C'est magnifique, mais ce n'est pas la guerre.'* Lucan was sent home by Raglan, and although never officially censured, was not employed again. Cardigan, whose cool courage made him a popular hero, became Inspector General of Cavalry after the war.

The Aftermath

Following the Charge, Raglan was still anxious to recover the redoubts, but was dissuaded from further action by the French commander, Canrobert. The battle was over. The Russians had lost 540 killed and wounded, the British 360, the French 38, and the Turks 260 men. In terms of lives lost Balaklava was little more than a skirmish. On 5 November, Menshikov launched his 'decisive' attack from the Inkerman, intending to drive the Allies into the sea. Despite the confusion and very heavy casualties, the Allies held on, but they would now have to face a winter in the Crimea. Tsar Nicholas anticipated a decisive victory by what were often described as Russia's most reliable generals, 'Generals January and February'! Nicholas died on 2 March 1855, from influenza.

Sevastopol finally fell to the Allies on 9 September 1855. By then the glory and drama of Balaklava had given way to an attritional struggle between entrenched armies, presaging the horrors of the Western Front in World War I. The Charge seemed like something from an earlier, romantic, age. Those who died that day escaped the terrible fate that would befall their comrades. Over the winter 1854–55 the British cavalry had been annihilated by cholera, dysentery, overwork, malnutrition, and exposure.

But within weeks of the Battle of Balaklava a popular myth had been created. The Charge was turned into a disaster to support contemporary political agitation, which wished to portray the aristocracy as the cause of Britain's military problems, as part of a wider campaign for administrative reform. The campaigners developed the image of stupid titled generals wilfully destroying working-class soldiers. Like all the best myths this version survived the original purpose for which it was created, and took on a life of its own. Other responses to the Charge, notably Alfred, Lord Tennyson's famous poem, celebrated bravery, but began the process of demeaning the battlefield achievement that has led many to see the Charge as a disaster. It may have been the result of a mistake in the transmission of orders, and costly, but it was not a disaster. The Charge gave the British cavalry a reputation for courage and discipline that will last as long as the history of war is studied.

ANDREW LAMBERT

Further Reading

Adkin, M., *The Charge* (London, 1997)

The Marquis of Anglesey, *A History of the British Cavalry: Volume II 1851–1871* (London, 1975)

Kinglake, A., *The Invasion of the Crimea: eight volumes* (London, Cabinet Edition, 1899)

Lambert, A., *The Crimean War: British Grand Strategy against Russia 1853–1856* (Manchester, 1990)

Lambert, A., and Badsey, S., *The War Correspondents: The Crimean War* (Gloucester, 1994)

Seaton, A., *The Crimean War: A Russian Chronicle* (London, 1977)

Gallipoli

25 April 1915

Too Many Commanders

'Damn the Dardanelles! They will be our grave!'

Admiral Jacky Fisher

It is a common military maxim that it is better for an army going into battle to have one plan and to try to stick to it than to try to change it at the last minute, or to have one mediocre commander in charge of the battle rather than two good ones who each try to take command. In trying to form a strategy after the first frenzied months of World War I, the British found themselves in exactly this position. The result was a battle which, although highly imaginative in its overall plan, failed through a catalogue of divided and conflicting ideas about who was in charge.

The Strategic Plan

By the end of 1914 the Western Front had become deadlocked. The narrow failure of the Germans to defeat the French and British quickly in August and September had been followed by a confused period of manoeuvre. This had left both sides temporarily exhausted, and unable to break through the enemy's hastily dug trenches, which stretched from the North Sea to the Swiss frontier. In October the Ottoman Turkish Empire (which stretched from modern Turkey to include most of the Arabian Peninsula) came into the war on the German side. Within the British decision-making elite, an 'Eastern' party began to emerge, which sought to make use of British seapower to outflank the stalemate in France and Belgium. A leading Easterner was Winston Churchill, the First Lord of the

Admiralty (the political head of the Royal Navy), who favoured an attack on the Dardanelles, the channel connecting the Aegean Sea with the Sea of Mamara. Little more than a mile wide at its narrowest point, 'The Narrows', the Dardanelles divided the Gallipoli peninsula in Europe from Asia Minor. There appeared to be enormous benefits if this area could be successfully attacked. Constantinople, the Ottoman capital, lay at the head of the Sea of Mamara, on the Turkish side of 'The Straits'. If a fleet could force the Dardanelles, Constantinople would be vulnerable to attack, and its capture would probably force the Turks out of the war. This would open up the Black Sea to Allied shipping, and thus revitalize the war effort of Britain's ally the Russian Empire, including the opportunity of exchanging Russian grain for British and French munitions. Germany and its ally Austria-Hungary would also be faced with a major new front in the Balkans, which would relieve pressure on the Western Front. At the time the opportunities seemed endless, although even with hindsight the prob-able outcome of a successful attack remains hard to assess.

Churchill's Cabinet colleagues did not all share his enthusiasm for an attack on the Dardanelles. Field Marshal Lord Kitchener, the Secretary of State for War (the political head of the Army, although technically still a serving soldier), refused to allocate ground forces to the expedition, as he believed that every man was needed on the Western Front. Thus on 13 January 1915, the British government's senior decision-making body for the war, the War Council, decided that 'The Admiralty should prepare for a naval expedition in February to bombard and take the Gallipoli Peninsula with Constantinople', a proposal that was confirmed on 28 January. The professional head of the Royal Navy, First Sea Lord Admiral Lord ('Jacky') Fisher, agreed to support the project despite his reservations. Vice Admiral S H Carden, commanding a British squadron in the Eastern Mediterranean, cautiously agreed that such an operation was feasible, but could not be hurried and might be costly. Kitchener's refusal to commit ground forces complicated the Royal Navy's task, since only troops could occupy Gallipoli and the Asia Minor coast, and thus neutralize their shore batteries.

The Naval Assault

This failure to coordinate a naval with a military attack was one of the most controversial issues in the entire campaign. Although a British naval squadron had rushed the Dardanelles in 1807, a joint military–naval report of 1906 had rejected the idea of such an exploit being repeated under modern conditions. In 1911 Churchill himself had declared that an attempt on the Dardanelles by ships alone would fail, and as recently as 8 January 1915 Kitchener had predicted that 150,000 troops would be needed to capture Gallipoli. Even if the fleet had forced the Straits, how could it take a land objective such as the Gallipoli peninsula without the support of troops? And how could it then 'take' the city of Constantinople? The answer appears to be that it was hoped that the arrival of an Allied fleet would cause Turkish resistance to collapse. Sir Edward Grey, the Foreign Secretary, admitted in February 1915 that, 'what we relied upon to open the Straits was a *coup d'etat* in Constantinople.' Such views were founded partly on the poor performance of Turkish forces in the Balkan Wars of 1912–13 and the initial campaigns of 1914. They were also based upon British notions of Turkish racial and cultural inferiority; the Turks were seen as 'Bashi-Bazouks' (a variety of tribesmen) who were usually easily beaten by the British in colonial wars. The initial plan for the Gallipoli campaign was thus the product of wishful thinking and underestimation of the enemy. The 1917 Dardanelles Commission of Enquiry aptly commented on the 'atmosphere of vagueness and want of precision which seems to have characterized the proceedings of the War Council'.

On 16 February 1915, British ships moved into the southern end of the Straits to commence minesweeping and the bombardment of the defending Ottoman forts. This was an entirely sensible move, judged from a purely naval perspective, but unfortunately it threw away one of the principal advantages the Allies possessed: surprise. In February there was only one Turkish division on the peninsula (9th Division), and twice, on 27 February and 3 March, parties of Royal Marines landed, encountering no real resistance. A *coup de main* launched by a small amphibious force would probably have succeeded in eliminating the forts overlooking the Dardanelles. Instead, the Turks did not waste the month's warning they were given of Allied intentions. The commander of Turkish Fifth Army, responsible for Gallipoli, was actually a German officer, General Otto Liman von Sanders, who set about reinforcing the area and strengthening their defences. By the time of the amphibious assault in April, the 9th Division had been joined on the peninsula by the 5th

Division, 7th Division, and 19th Division, with the 3rd Division and 11th Division on the Asiatic shore.

The blue waters of the Dardanelles witnessed an awesome sight on 18 March 1915. Fourteen British and four French battleships deployed in three lines, with accompanying craft, sailed up the Straits determined to force a passage through to Constantinople. At 11.30 a.m. the heavy guns of the battleships in the van began to pound the shore forts at long range. But Vice Admiral Sir John de Robeck, the force commander (Carden having fallen sick), was to find that the Turkish defences were tougher than he supposed. The shore batteries had been reinforced by 24 mobile howitzer batteries, 11 lines of contact mines (a total of 344 mines) had been laid, and three 18-inch torpedo tubes had been mounted in place. By early afternoon, the Allied fleet had begun to take casualties. The French ships *Gaulois* and *Suffren* were badly damaged by shellfire, and the *Bouvet* exploded and sank with 600 deaths. HMS *Inflexible* hit a mine, *Irresistible* was torpedoed and sank, and *Ocean* suffered the double blow of striking a mine and being hit by a shell, and also sank. Accepting defeat, de Robeck broke off the engagement, leaving the minefields of the Dardanelles substantially intact. This fact is by itself enough to cast doubt on the popular notion that a fresh attack on 19 March would have broken through the Narrows. The Turkish guns had fired off much of their ammunition, but it seems that they had enough to keep the minesweepers at bay.

If nothing else was learned from the debacle of 18 March, the lesson that battleships could not expect to survive in confined waters sown with unswept mines had been hammered home. If an amphibious force had been simultaneously landed on the Gallipoli peninsula, and had succeeded in capturing the forts from the rear, the minesweepers could have carried out their role unhindered. Ironically, by the time of the naval repulse on 18 March, such a joint operation had become a real possibility.

The Plans for the Allied Landing

Even as the British battleships began to pound the forts on 16 February, Lord Kitchener changed his mind: he could, after all, spare troops to send to the Dardanelles. The commander of the 'Mediterranean Expeditionary Force' or MEF, General Sir Ian Hamilton, was not however appointed until 12 March, and left Britain next evening. Major General Walter Braithwaite, appointed as Hamilton's chief of staff (although Hamilton had not requested him) was given only 48 hours to create a staff. Similarly, a junior staff officer had a day in which to calculate the supply requirements of the force. This last minute mad scramble was all too typical of the British approach to the Gallipoli campaign. Hamilton had 75,000 men, just half of the number Kitchener himself had said was necessary to capture Gallipoli. These included some French colonial troops, the self-named 'Incomparable' 29th Division, which was effectively the last of the old pre-war regular British Army, the Royal Naval Division (RND) formed by Churchill from volunteers, sailors, and marines, and the volunteer Australian and New Zealand Army Corps (ANZAC), which in the spring of 1915 was an unknown quantity. Hamilton had no strategic surprise, little useful intelligence, no purpose-built landing craft, and few aircraft. His divisions were short of artillery and the logistic (supply) system was in chaos.

Perhaps most serious of all, Hamilton and de Robeck had fundamentally different concepts of operations. Hamilton thought that his landing would be accompanied by another attack by warships; and that de Robeck would open up another front once the troops had secured their beachhead. His naval counterpart, by contrast, wanted Hamilton to capture the Gallipoli peninsula before he would commit his vessels to the perils of the Narrows, something of which Hamilton was unaware. The fatal weakness of the divided command was already becoming apparent.

Hamilton was faced with the unenviable task of attempting the first ever opposed amphibious landing against troops armed with modern weapons. The assault was to take place on 25 April 1915. There were to be two main landings: Major General Sir Aylmer Hunter-Weston's 29th Division was to land on the southern tip of the Gallipoli peninsula at Cape Helles; the ANZACs under Major General William Birdwood were to land a mile north of Gaba Tepe, on the western coast of the peninsula. French forces were to demonstrate by making a diversionary landing at Kum Kale, on the coast of Asia Minor. Major General A Paris's Royal Naval Division was to sail up

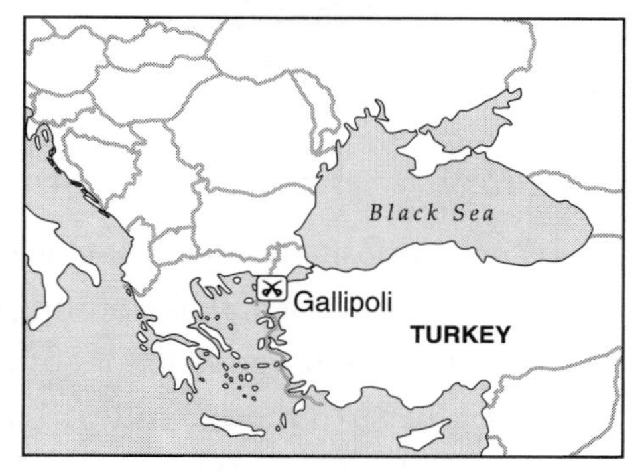

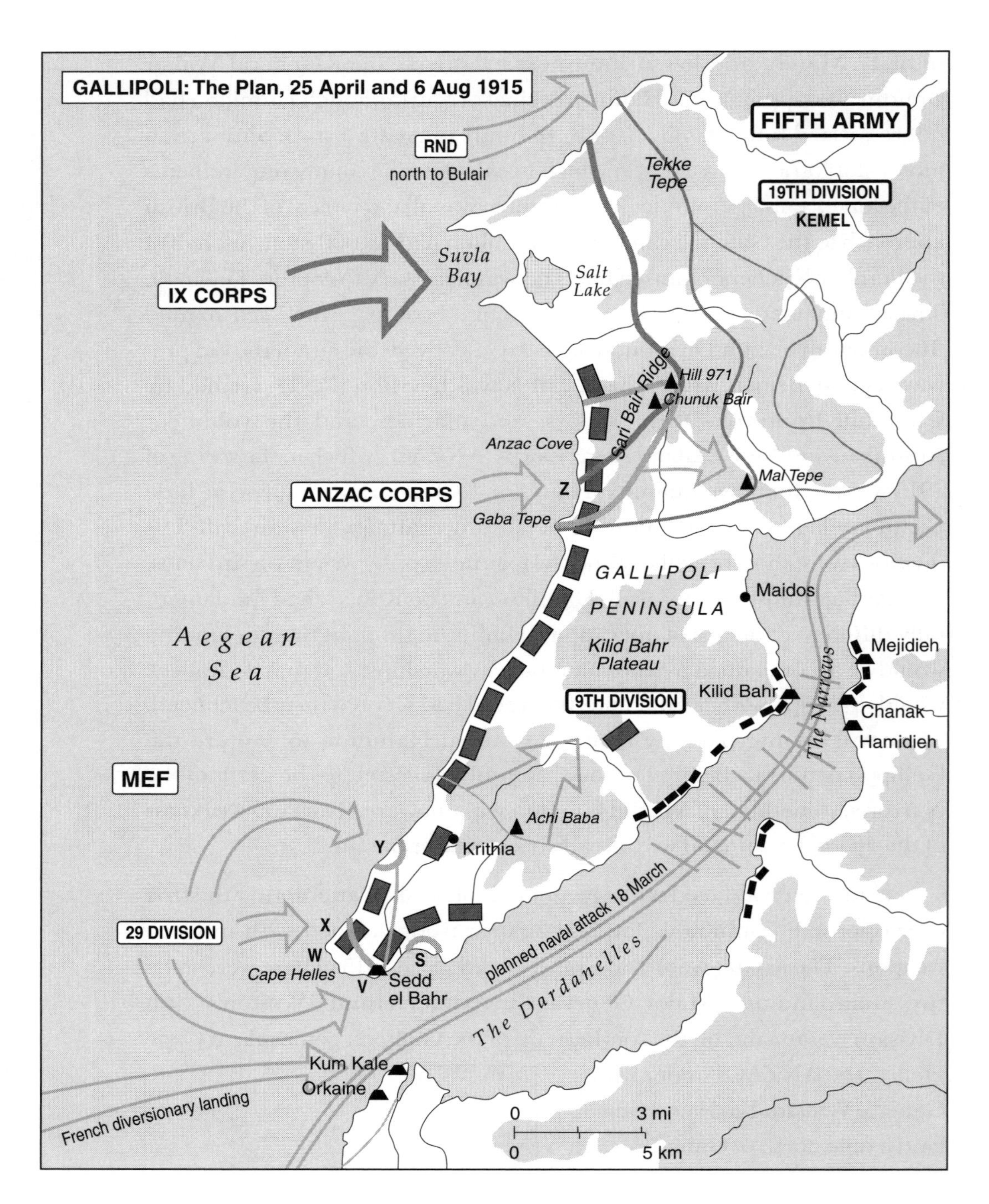

and down opposite Bulair, at the northern end of the peninsula, threatening a landing but not actually putting troops ashore.

The Gallipoli peninsula is some 50 miles long, but its width varies from 12 miles to a mere 3 miles. The terrain is very unpromising for an attacker: hilly and broken, it hands many advantages to the defender.

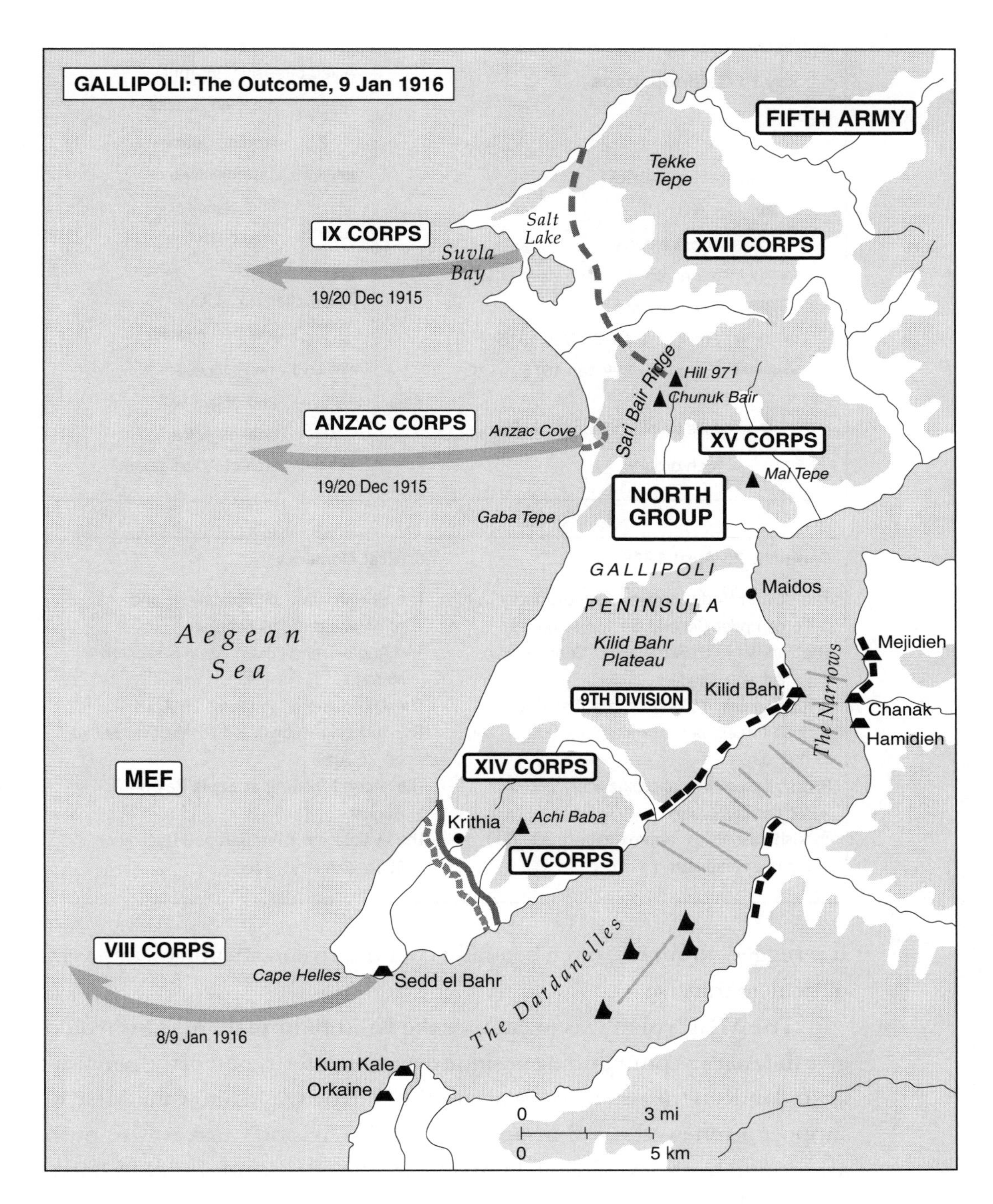

Three major topographical features dominate the southern part of the peninsula. The hill of Achi Baba (700 ft) rises above the Helles area. About the same height is Kilid Bahr plateau, some four miles to the northeast of Achi Baba. The Sari Bair ridge (1,000 ft) is about ten miles to the north of Achi Baba, near what was shortly to become immortalized as 'Anzac Cove'.

key to Gallipoli maps	Allied offensive 25 April
	planned attacks
	Z landing beach
Turkish	1st objective
fort	2nd objective
shore battery	final objective
minefield	
deployments	Allied offensive 6 Aug
Front Line 19–20 Dec 1915	planned attacks
Front Line 8–9 Jan 1916	1st objective
	2nd objective
Allied sinking March 1915	final objective
high ground	limit of Allied gains

Gallipoli, 25 April 1915

The British Mediterranean Expeditionary Force under General Sir Ian Hamilton

The Turkish Fifth Army under General Otto Liman von Sanders

British forces: 75,000 on the first day

Turkish forces: approximately 42,000 on the first day

British casualties: approximately 265,000 for the campaign

Turkish casualties: approximately 300,000 for the campaign

Critical Moments

The British naval bombardment and minesweeping 16 February

The Anglo-French main naval attack 18 March

The Anglo-French landings 25 April

The Turkish response led by Mustafa Kemel on 25 April

The second landing at Suvla Bay on 6 August

The evacuation from Gallipoli December 1915–January 1916

It is rugged, dramatic, often beautiful country, riven with gullies, and very difficult to traverse.

The MEF's plan was to capture the Kilid Bahr plateau. This would give the Allies a commanding position over the eastern coast of the peninsula, and indeed the Asiatic coast across the Narrows, and allow the MEF to support a renewed naval assault. The 29th Division's task was to push ashore and by the evening of 25 April seize the vital ground of Achi Baba. Birdwood's ANZACs were to advance about five miles to the east of their landing at Anzac Cove, and take a ridge around Mal Tepe hill. Unfortunately, Hamilton and his staff seemed to assume that the landing was going to be the most difficult part. Thus while the orders for the assault were highly detailed, the orders for the exploitation were vague. Little emphasis was placed on the importance of quickly taking Achi Baba.

Was this the best available plan? Certainly, most of Hamilton's senior commanders were at best lukewarm towards it. Given that the crucial element of surprise had been sacrificed, should the MEF have attacked Gallipoli at all? Hamilton was in fact hemmed in by his orders. Kitchener had forbidden extensive operations on the Asian coast. Indeed, in early April Hamilton stated that, 'I have no roving commission to conquer Asia Minor ... I have not come here for any purpose whatever but to help the Fleet through the Dardanelles.' Hamilton, a friend and protégé of Kitchener's, was on this issue as on virtually every other, loath to stand up to him. The strength of the Turkish defences at Bulair made the option of an attack in the north unattractive, while the cramped conditions at Helles risked the attackers being sealed off in their beachhead by the Turks. The main attraction of Helles was that the navies would be able to offer close support during the landing and then in the advance inland. In the event, naval support of the landing was to be mostly fairly ineffective.

The Landings

The amphibious assault commenced at dawn on 25 April 1915, to be forever after celebrated in the Antipodes as 'Anzac Day'. The 1st Australian Division was put ashore about one mile north of its original objective, either through an error or a last minute change of plan. Initially, the Turks put up little resistance, but as the inexperienced ANZACs moved inland, climbing a steep cliff and moving through the rough terrain, their advance lost all semblance of order. Although a handful of Australians reached Chunuk Bair, the Turks began to mount improvised counterattacks and push the ANZACs back. The leadership of Lieutenant Colonel Mustafa Kemel, the commander of the Turkish 19th Division (and the post-war President of the new Turkish Republic Kemel Atatürk) was crucial. If not for Kemel's intervention, the Australian advance, ragged as it was, might have prevailed. As it was, Kemel's action bought valuable time for the Turks to regroup and for their reserves to arrive. By the evening of 25 April, the Turks were driving forward remorselessly, and the Australians were pinned back on to the beach and the heights above Anzac Cove. The ANZACs had got 15,000 men ashore during the day, suffering 2,000 casualties, but the chance of winning the ground campaign had quickly vanished.

Elsewhere, Allied fortunes were mixed. At Kum Kale, the French diversionary landing was a success, and to the north at Bulair the feint of the RND tied down Turkish reserves. At Helles, the British regulars of 29th Division faced light opposition at the rugged cliff-like landing areas optimistically designated S, X, and Y beaches, but at V and W beaches the Turks defended fiercely. At V beach an old collier, *SS River Clyde* was deliberately run aground, and two battalions (2nd Hampshires and 1st Royal Munster Fusiliers) emerged from sally ports cut in the side of the ship. Unfortunately, the water was too deep, and many men were drowned or hit by Turkish fire. Another Irish unit, 1st Royal Dublin Fusiliers, was landed more conventionally from boats, but they too were held up at the water's edge. The survivors of the attacking force, wounded and unwounded alike, remained huddled behind the scant cover afforded by sand on the beach.

At W beach, in another bloody assault, the 1st Lancashire Fusiliers were busy winning 'six VCs [Victoria Crosses] before breakfast'. Here, the Turkish defenders coolly held their fire until the British boats were only a hundred yards from the beach, and then raked the ranks of the attackers. As at V beach, the British infantry struggled to find a way ashore. The brigade commander, in a boat slightly behind the first wave, spotted an opportunity to land out to one flank, which was apparently undefended. He succeeded in leading a small party into the rear of the Turkish positions, although he was wounded in the process. By nightfall, the British were ashore at both V and W beaches, although at great cost. Achi Baba and the Straits were still tantalizingly out of reach.

At Y beach, the British were handed the best opportunity of the day to reach their final objectives. Two battalions landed to find the Turks were nowhere to be seen. Unfortunately, instead of striking out boldly to take the Turks in the rear, the British troops, aided by unclear orders and confusion as to who exactly was in command, stayed put. Worse, Hunter-Weston became obsessed with the battle for V and W beaches and failed to take the opportunity to reinforce Y beach . Hamilton, from his headquarters on board the battleship *HMS Queen Elizabeth*, realized Hunter-Weston's error but only advised his subordinate to change his plans rather than giving him a firm order. Hunter-Weston ignored Hamilton's advice, and the fleeting moment for decisive action passed.

Stalemate and Failure

The morning of 26 April dawned with the Allies firmly ashore at both Anzac and Helles, but with the original objectives as far away as ever. Instead, the Allies were committed to the very type of trench warfare the whole Dardanelles campaign was supposed to avoid. A gruelling series of indecisive attritional battles ensued in May and June, with the Allies inching closer to their objectives without ever looking likely to achieve them. On 6 August a major new phase of the campaign began when a fresh force was landed in a surprise attack at Suvla Bay, three miles north of Anzac. This, coupled with a renewed attack from Anzac, briefly came close to success, but then also ended in disappointment. Many of the weaknesses in British command revealed by the assault of 25 April were all too apparent in this second attack. After further trench fighting, the Allies evacuated the Gallipoli peninsula in two stages, Anzac and Suvla on 19–20 December, and Helles on the night of 8–9 January 1916. Unlike the Allied landings, the evacuations were a complete success.

Why was there such a great disparity between what was supposed to happen at Gallipoli and what actually did happen? Numerous factors could be considered, ranging from lack of adequate intelligence to logistic confusion, but there is little doubt that failures of cooperation contributed mightily to the end result. The original divorce between the naval assault and the ground forces was disastrous. A joint naval–military attack, launched early enough to gain surprise, might have succeeded in getting the Fleet through the Straits – although that is not to argue that Constantinople would then have fallen, or that if it had, Turkey would have dropped out of the war. Even after the failure of the naval attack of 18 March, close cooperation between the naval and mili-tary authorities could have improved on the performance of 25 April. In short, the lack of a proper British amphibious doctrine and capability fatally undermined the entire Dardanelles campaign.

G D Sheffield

Further Reading

Aspinall-Oglander, C.F., *Military Operations Gallipoli, Vol.1* (London, 1929) [British official history]

Rhodes James, R., *Gallipoli* (London, 1965)

Steel, N. and Hart, P., *Defeat at Gallipoli* (London, 1994)

Mission Impossible

'The plan was smooth on paper, only they forgot about the ravines.' Russian military proverb

If a misunderstanding can ruin a good plan and produce a disaster, often far more costly and more tragic is a plan for a battle that is based on some completely false premise. The army finds itself trying to do something which simply cannot be done, because somewhere in the planning process either a critical point was missed, or inconvenient facts were quietly disregarded, or the information used to produce the plan was missing a detail. It is only after the battle starts that such plans are suddenly shown to be unworkable, and by then it is usually too late.

Kunersdorf

12 August 1759

The Failure of Reconnaissance

'I shall not survive this cruel misfortune. The consequences will be worse than defeat itself. I have no resources left, and, to speak quite frankly I believe everything is lost. I shall not outlive the downfall of my country. Farewell, forever!' Frederick the Great

One of the most common sources of failure in military planning comes from not knowing what, as the Duke of Wellington once put it, is 'on the other side of the hill'. Reconnaissance, the discovery of what the ground for the battle is like and where the enemy might be, has always played a critical role in warfare, and without it even the most talented of commanders has been known to come to grief. As this uncharacteristically poor performance by the most able of 18th-century Europe's soldier-kings shows, the other side of the hill can sometimes contain a nasty surprise.

Frederick the Great's Prussia

In the course of the War of the Austrian Succession (1740–48) and the Seven Years' War (1756–63), the Kingdom of Brandenburg-Prussia in northern Germany emerged as one of the major European powers, alongside Great Britain, Russia, Austria, and France. The rise of Brandenburg-Prussia in the middle of the 18th century was principally the result of two factors. First, Prussia's highly efficient state administration allowed this relatively poor country, with a population of under two and a half million people, to raise an army of 154,000 troops, prompting the French philosopher Voltaire to comment that, 'Prussia is not a state with an army. It is an army with a state.' The size of the Prussian Army, and the professionalism of its officers and men made it one of the finest in Europe, and capable of challenging the military forces of the great powers. But, above all else, the rise of Prussia was also inextricably linked to the outstanding abilities of Frederick II (1712–86) who succeeded as King of Brandenburg-Prussia in 1740 and who became known as Frederick the Great (or, less flatteringly, as 'Old Fritz' by his soldiers).

Frederick was a highly intelligent, skilful, ruthless, and aggressive military and political leader, who used the superb army he inherited from his father to wage a series of wars that advanced Prussia into the front rank of European powers. At times the odds were stacked so heavily against Prussia that it seemed inevitable that she would be overwhelmed by her stronger and more resilient French, Austrian, and Russian enemies. Only Frederick's brilliant generalship, and obstinate refusal to accept defeat, averted disaster. However, the failure of reconnaissance leading to the catastrophic defeat at Kunersdorf seemed for a time to spell the end not only for Frederick but also for Prussia itself.

The Seven Years' War

The Seven Years' War began in August 1756 when Frederick launched a pre-emptive invasion of Saxony, in response to Austria and Russia's signing of an alliance which he quite correctly believed was the prelude to an attack upon Prussia. Empress Maria Theresa of Austria wished to crush the Prussian challenge to Austria's position as the principal German state, and to recover the province of Silesia, seized by Frederick during the War of the Austrian Succession. The Russian Empress Elizabeth, meanwhile, wanted to remove any Prussian opposition to her aim of dominating eastern Europe. By 1757, France, Sweden, and most of the other minor German states had sided in this alliance against Prussia, and Frederick found himself outnumbered and threatened on all sides. His only ally was Great Britain which, led by Prime Minister William Pitt ('the Elder'), was already at war with France.

In the opening years of the war, Frederick's fortunes fluctuated dramatically. In 1757 victory was secured by only the narrowest of margins at Prague on 6 May, whilst at Kolin on 18 June the Prussians suffered a resounding defeat. But Frederick then went on to win two spectacular victories at Rossbach on 5 November and Leuthen on 5 December. In 1758 the Prussian Army fought the Russians in a bloody and indecisive battle at Zorndorf on 25 August, and suffered a heavy reverse at Hockhirch on 14 October. Although in these first years Frederick managed to keep his foes at bay, in the process of so doing his army and resources were steadily drained away without any prospect of victory.

The losses in men and *materiel* (military equipment and stores) during the previous two years, coupled to the fact that Prussia was facing invasion from three directions – the French from the west, the Austrians from the south, and the Russians from the east – meant that in 1759, for the first time in the war, Frederick was forced to adopt a more defensive strategy, and let the enemy carry the war to him. He accordingly divided his army up to meet each threat. His brother Prince Henry (an intelligent and cool-headed general) covered the west with an army in Saxony, whilst General Henri-Auguste de la Motte-Fouqué covered Upper Silesia against an Austrian advance from the south, and General Carl von Dohna guarded against the Russians in Poland. Frederick with the main army of 44,000 troops was stationed in the centre in Lower Silesia, from where he could move to support the other Prussian forces against a major attack by any of the three Allies.

Spring 1759 saw little movement on the part of the main armies. Although Prince Henry and General Fouqué succeeded in destroying the supply depots of the Allied forces facing each of them, the Austrians did not react and patiently continued to concentrate their main forces at Landshut. His opponent's behaviour was extremely annoying for Frederick, whose natural temperament was for action, and who wrote that he was desperately pinning his hopes for that year's campaign upon 'a good decisive battle, which will render it safe to send detachments to where the need is most urgent'. His frustration increased in May, when the Austrian Army under Field Marshal Leopold Daun left its quarters at Landshut, but instead of offering battle the Austrian commander was content to put his troops through a series of manoeuvres.

The Campaign Begins

In contrast to the first six months of the year, July and August 1759 saw feverish activity by the opposing armies, which was to culminate on 12 August with the battle Frederick desired. In mid-July the Russian army under Field Marshal P S Saltykov advanced to the River Oder and occupied the town of Frankfurt-am-Oder, only about 50 miles from Berlin, defeating Dohna's forces on the way. With the door to Berlin wide open, Frederick quickly gathered 19,000 troops and began a series of forced marches northwards to block any further Russian advance. In the blazing heat of late summer, Prussian columns snaked across northern Silesia, throwing up huge clouds of dust. At the same time the Austrian commander Daun dispatched two of his subordinates, Major General G E Loudon with 24,000 troops and Lieutenant General A Haddik with 17,300 troops, to join up with Saltykov's force.

Frederick moved his troops at a relentless pace in order to prevent the junction of Austrian and Russian forces. Although he successfully forced Haddik to abandon his march northwards by destroying his baggage on 2 August, the Prussians failed to overtake Loudon, who joined Saltykov at Frankfurt on 5 August, giving the Allies a combined strength of 64,000 troops. Next day, Frederick united with Dohna's shattered troops and set about restoring their morale. The concentration of Prussian forces on the left bank of the Oder was completed by the arrival of Lieutenant General F A von Finck's corps from Berlin, giving Frederick an army of 49,000 troops.

The King resolved to cross the Oder and attack the Russian–Austrian Army without delay. A decisive victory would not only remove the immediate threat to Berlin posed by Saltykov and Loudon's forces, but release the Prussian forces facing them for an advance southwards to attack Daun's army. With both of the main Allied armies defeated, Frederick would be free to invade Austrian territory and try to pressure Maria Theresa into making peace. At the very least he would be in a far better position to prosecute the war in 1760.

During the early hours of 10 August an advance guard of a regiment of Prussian fusiliers was ferried across the Oder near Custrin, and established a bridgehead behind which two bridges and a pontoon were constructed during the day. In the late evening the infantry and artillery of the

Prussian Army crossed over to the right bank of the Oder, whilst the cavalry waded the river in the shallow water near Otscher. The baggage was left at the bridges under the protection of a force of infantry and cavalry commanded by Major General Johann Jacob Wunsch, who had orders that if the opportunity arose he was to recapture Frankfurt.

The Prussian Reconnaissance

Next day was 11 August, and Frederick attempted to gain what information he could about the disposition of the enemy and the nature of the surrounding terrain. In the afternoon he rode forward to the hill of the Trettiner-Spitzberg, in front of his advance guard, to personally view the enemy positions to the south. A careful study revealed that Saltykov and Loudon had taken up a formidable position, one that made a frontal attack not simply difficult, but more than likely suicidal.

The Allies had entrenched themselves along a series of low sandy hills stretching away southwest towards marshy ground beside the Oder, which therefore excluded any attempt to outflank them in this direction. To the front of their position was the Huhner-Fliess, a marsh that could only be traversed by two walkways, and clearly any attempt to cross this bog would be painfully slow and expose the troops to heavy and prolonged fire. Little intelligence could be gained about the type of terrain behind the Allies, around the village of Kunersdorf. Only two people with Frederick's army were familiar with the area, but Major Linden's hunting excursions had not furnished him with the sufficient information to give a tactical assessment of the ground, and a local forestry official proved useless because he was struck dumb at the sight of the king, who, despite adopting a gentle manner, was unable to coax him into speaking.

Despite the lack of information about what lay to the rear of the Allied position, Frederick decided to forgo any further reconnaissance and attack immediately. Lieutenant General Finck with 8 battalions of infantry and Lieutenant General Scharlmer's 40 squadrons of cavalry were to

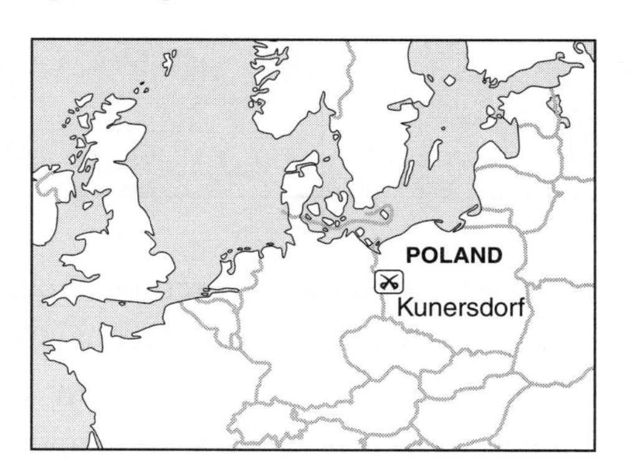

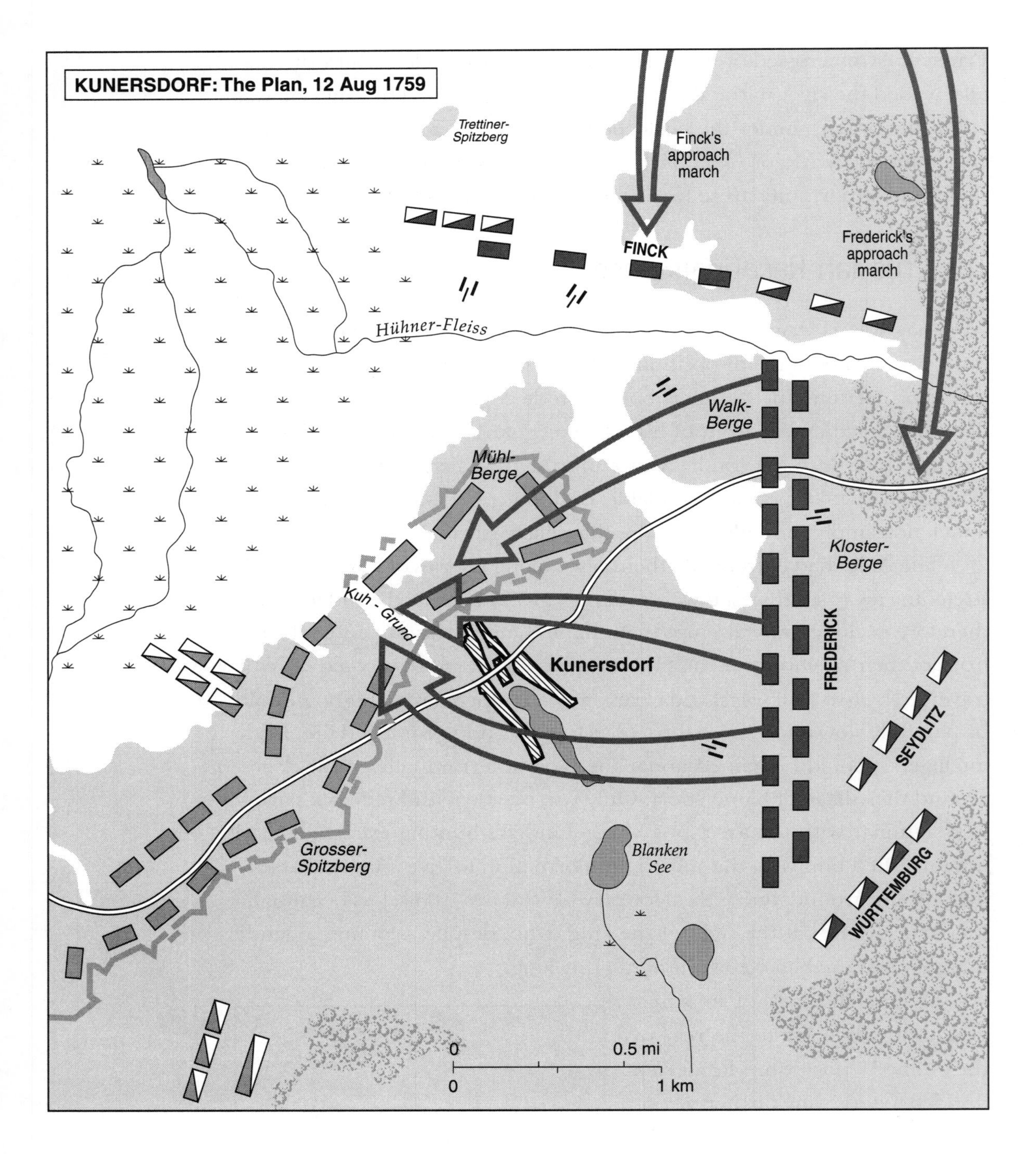

demonstrate along the Trettiner-Spitzberg in order to convince the Allies that Frederick intended to mount a frontal assault. Then in a repeat of the manoeuvre he had carried out against the Russians at Zorndorf, Frederick and the bulk of his army would undertake a long night march

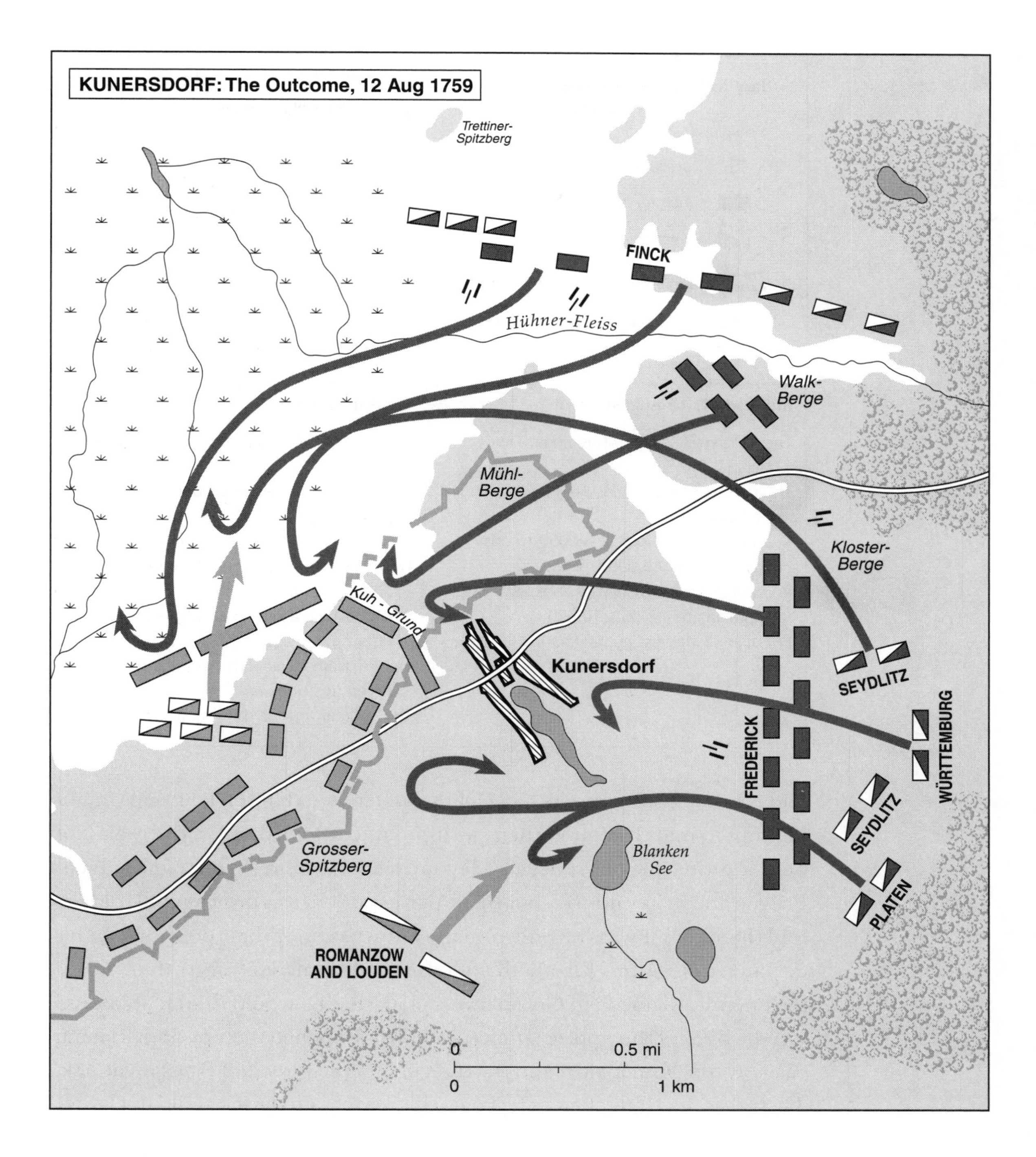

through the Neuendorfer Heich woods to the east of Kunersdorf, to come around the Allies' flank and attack their exposed and unentrenched rear.

On the other side, Saltykov and Loudon planned to remain in their strong positions and await the expected Prussian attack. But the Austrian

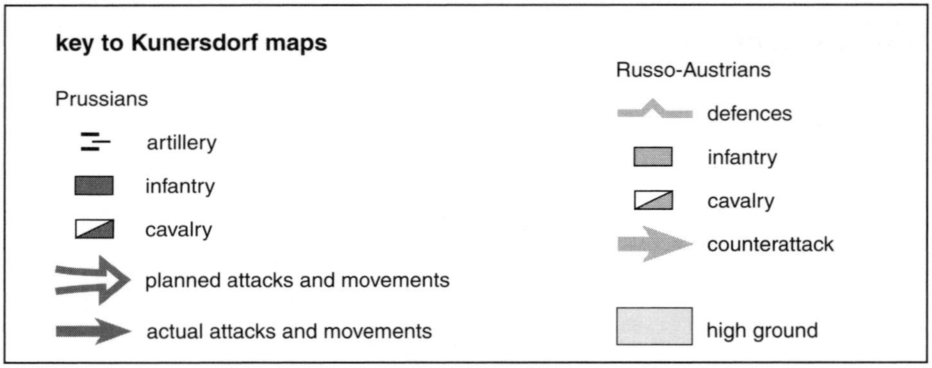

key to Kunersdorf maps

Prussians

- artillery
- infantry
- cavalry
- planned attacks and movements
- actual attacks and movements

Russo-Austrians

- defences
- infantry
- cavalry
- counterattack
- high ground

Kunersdorf, 12 August 1759	**Critical Moments**
The Russian–Austrian Army under Field Marshal P S Saltykov	The Prussian flank march before dawn on 12 August
The Prussian Army under King Frederick the Great	Frederick's deployment problems east of Kunersdorf
Russian–Austrian forces: 64,000 troops	The Prussians capture the Mühl-Berge
Prussian forces: 49,000 troops	Frederick attacks the new Allied positions on the Kuh-Grund fails
Russian–Austrian casualties: unknown but approximately 7,000–10,000	Finck's attack across the Hühner-Fleiss fails
Prussian casualties: approximately 19,000	The Prussian cavalry charges against the Allied positions fail
	The general rout of the Prussian forces

and Russian commanders had clearly anticipated that the Prussians might attempt to march around their positions and attack them in the rear. Not only was this one of Frederick's favourite manoeuvres – he had used it at Prague and at Zorndorf – but their own experience as professional soldiers told them that it was the only practical way to attack their positions. In his dispatch to Empress Elizabeth after the battle, Saltykov wrote that, 'when the enemy advanced to Geritz it was plain that he would march about us.' On the basis of this appreciation they had entrenched their positions facing east towards Kunersdorf and the direction of the intended Prussian attack. This was something that, as he began his march, Frederick could not know.

The Battle of Kunersdorf

Around 2.00 a.m. on the 12 August, the Prussian Army began to move to its left in two columns. The first line of infantry was preceded by

Lieutenant General Frederick William von Seydlitz and his left wing of cavalry, while the cavalry of the right wing, commanded by Lieutenant General the Prince of Württemburg, led the second column through the woods. Progress along the overgrown, muddy tracks was slow and tiring, but by daylight the army emerged from the woods, and 45 battalions of infantry were drawn up in two lines with the cavalry forming a third behind them.

It was now that Frederick's impatience to attack the Russian–Austrian Army before carrying out a full reconnaissance of their positions near Kunersdorf proved to have been a grave mistake. Looking west towards the Allied lines it was clear that the positions facing the Prussian Army were more heavily fortified than those to the northwest. Furthermore, when the Prussian Army tried to extend its left flank, it was discovered that this was prevented by a series of ponds extending south of the village, of which the largest was the Blanken See, and this cramped its deployment. Instead of a broad front attack against the exposed rear of the Allied Army, Frederick would be forced to attack strong defences along a very narrow frontage.

Rather than call off or postpone the attack, Frederick decided to re-align his army and launch his main attack against the northernmost Russian positions on the Mühl-Berge. There followed a period of delay, as the gunners moved their pieces through the woods into new positions, and the lead regiments of the infantry columns retraced some of their steps northwards. The troops, who were already tired from several days of hard marching and the lack of food and sleep during the previous night, now suffered heavily in the oppressively hot weather.

At 11.30 a.m. around 60 Prussian guns opened up a heavy and sustained fire against the Mühl-Berge. After an hour, with the Russian artillery beaten into silence, Frederick ordered Major General Schenkendorf's and Major General Linstedt's brigades of his advance guard to storm the position. The shattered regiments of the Russian Observation Corps on the Mühl-Berge offered little resistance as the Prussian grenadiers surged over the entrenchments in perfect order, capturing over 80 enemy cannon. Once on the Mühl-Berge, the Prussian troops deployed out into line and drove back four Russian regiments which Prince Gallitzin had brought up to reinforce his troops on the heights.

Frederick's Fatal Decision

Several of Frederick's senior officers, among them Finck and Seydlitz, now intervened to argue that with the capture of the Mühl-Berge there was no need to continue the attack, because with their line compromised the Allies would be forced to abandon their camp and withdraw during the night. Frederick, however, was unwilling to countenance caution, and with the Russian left in disarray he was keen to push forwards and complete the destruction of the rest of the Allied forces. There followed a delay, while the artillery batteries were moved on to the top of the sandy heights, and the infantry regiments from the Prussian right moved up to reinforce the grenadiers.

During this pause, the Allies also took the opportunity to reform their line along a steep-sided valley called the Kuh-Grund which bisected the low range of hills on which they were deployed from east to west, and isolated the Mühl-Berge from the rest of the Russian–Austrian Army to the south. Saltykov ordered Generals Panin and Campitelli to move their brigades from the southern part of his line to form a line along the Kuh-Grund, while on his own initiative Loudon brought up several more regiments from the unengaged right flank.

In making this redeployment, because the constricted nature of the front between Kunersdorf and the swamp of the Hühner-Fleiss meant that no more than two regiments could be deployed in a line, the Allies were forced to draw up their forces in six or seven lines. Frederick's infantry was about to attack strong Allied forces deployed in depth along the natural defensive line provided by the Kuh-Grund, whose existence was quite unknown to Frederick.

Once their preparations were complete, the Prussians renewed their attack under the cover of heavy artillery fire. The grenadiers of the advance guard supported by the infantry of the right flank pushed against Panin's line along the Kuh-Grund. The steep sides of the slope disordered the Prussians, and despite heroic efforts they were unable to drive the Allies back. These attacks were supported by infantry from the centre and left of the Prussian Army, who marched around to the north of Kunersdorf, and attacked into the southwestern side of the Allied position, also with little success. As the day wore on the Prussian infantry, thirsty and tired from the heat, made repeated attempts to break the enemy line.

But with their centre and right unengaged, the Allies were able to feed fresh troops from these positions into the battle and stem the Prussian onslaught.

In an effort to break the stalemate, Finck brought his corps across the Huhner-Fleiss to engage the left of the Russian–Austrian line along the Kuh-Grund. His eight battalions struggled across the swamp, taking heavy casualties from the fire of Austrian cannon and Russian howitzers on the high ground. Despite repeated attempts, his men made no progress. About the same time some of the Prussian cavalry mounted a further abortive attack against the northwest flank of the Allied line.

As the day wore on, Prussian attempts to crack the Allied line became increasingly desperate. In particular, the cavalry was called upon to carry out virtually impossible tasks, and Frederick ordered Seydlitz to bring forward part of his cavalry and attack enemy troops around the Kuh-Grund. The cavalry debouched through broken terrain to the left of Kunersdorf, and after forming under heavy enemy fire advanced. Despite initial success the appearance of three new Russian regiments drove them back. With Seydlitz wounded soon after, the Prince of Württemburg assumed overall command of the cavalry, but his attempt to breach the Allied line proved as unsuccessful as the first attack when his men fled rather than obey orders.

By 4.00 p.m. the Prussian infantry were too exhausted to mount a new advance, and the only fresh forces were those cavalry regiments which had not yet seen action. With both Seydlitz and now Württemburg wounded, command of this force fell to the rash Lieutenant General D F von Platen, who without orders from Frederick decided to attack the Russian–Austrian entrenchments to the south of Kunersdorf. The lead Prussian regiment, the Schorleus Dragoons, dashed forwards in a futile attack against the heavily defended bastion of the Grosser-Spitzberg, and were annihilated by the Russian guns. Meanwhile, as the rest of Platen's regiments filtered between the ponds around the village, Count Romanzow and Major General Loudon led forward the combined Russian and Austrian cavalry, which hit the Prussians as they struggled to deploy. Caught unprepared and driven back on to the ponds, the Prussian cavalry were unable to resist the onslaught of the Allied horse, and were soon routed.

The End of the Battle

The sight of their cavalry fleeing broke the morale of the exhausted and battered Prussian infantry, who fell back in disorder until a general rout ensued. Despite the most strenuous efforts Frederick was unable to stop the panic, and that night when he reached the bridges across the Oder, the army could muster only 3,000 troops from an original strength of 49,000. Over the next few days another 15,000 men rejoined their regiments, but the army was shattered as a fighting force, and incapable of preventing an Allied advance on Berlin. Convinced that he had lost the war, Frederick sunk into a deep depression, lamenting the fact that he had not been killed during the battle.

Prussia and Frederick were only saved from total defeat by what became known as 'the miracle of the House of Brandenburg'. Instead of following up their success in battle and crowning it with victory over Prussia, the Austrian and Russian commanders quarrelled over their next moves, until the onset of winter forced them to retire. Frederick was indeed fortunate to survive the debacle of Kunersdorf. His over-eagerness for battle led him to disregard a careful reconnaissance of his oppon- ents' positions, and presented his tired army with an impossible task. Even so, it should be remembered that in spite of his errors on 12 August 1759, Frederick was the foremost commander of his age, and one of the greatest in history. Napoleon Bonaparte paid tribute to Frederick's genius after his own defeat of Prussia in 1806, when, standing before the king's tomb, he turned to his generals with the words, 'Hats off gentlemen! If he were still alive we would not be here.'

TIM BEAN

Further Reading

Black, J., *European Warfare 1660–1815* (London, 1994)

Duffy, C., *Frederick the Great: A military life* (London, 1985)

Duffy, C., *The Army of Frederick the Great* (London, 1974)

Showalter, D., *The Wars of Frederick the Great* (London, 1992)

The First Day of the Somme

1 July 1916

The Great Illusion

'You are about to attack the enemy with far greater numbers than he can oppose to you, supported by a huge number of guns... You are about to fight in one of the greatest battles in the world, and in the most just cause... Keep your heads, do your duty, and you will utterly defeat the enemy.' Special Order of the Day to the British 94th Brigade

If some battle plans go almost perfectly right, there are others that produce such complete disaster that afterwards it is hard to reconstruct the degree of self-delusion present in the minds of the generals who drew them up. It almost seems as if they sat down and deliberately worked out how to get their own men killed to no purpose. Of all the battles that fit this picture, the first day of the Battle of the Somme, 1 July 1916, has gone down in history as one of the greatest military disasters of all time. Although in some areas limited gains were achieved, the first day of the battle achieved far less than had been hoped. In 1916, with World War I already two years old, the mass British volunteers of 1914 were starting to arrive in large numbers on the still deadlocked Western Front. It has been argued that prior to the opening of the battle some commanders feared that this new and largely volunteer British Expeditionary Force (BEF) lacked the training and experience necessary to make the attack a success. However, although this was the new BEF's first major test, the blame for the debacle on 1 July does not lie with the volunteer soldiers, but with the fatally flawed and unrealistic plan devised by the professionals who led them.

The British Battle Plan

The detailed planning of the Somme offensive (named after the region of France, the River Somme itself flowing through the French positions just to the south) fell to the Fourth Army commander, Lieutenant General Sir Henry Rawlinson. Fourth Army was to make up the bulk of the BEF troops that were to advance on an 18-mile front on the first day of the battle, and a complex plan was meticulously pieced together for them. Having studied German defences in the Somme area, which invariably consisted of numerous heavily fortified positions situated on high ground, Rawlinson planned for a lengthy preliminary artillery bombardment followed by an infantry advance towards shallow object-ives, and then a period of consolidation before the start of the next phase of operations.

Central to the infantry plan was a clear appreciation of what the artillery could realistically achieve. Having carefully calculated the number and types of guns at his disposal, Rawlinson chose initial objectives that often comprised of just the German front line and some tactical positions beyond, resulting in a plan for an infantry advance of some 1,000 to 2,000 yards in order to keep the dilution of his artillery fire to a minimum. The importance of the guns to Rawlinson's offensive can be seen in his words to a colleague in February, '[The Somme] is capital country in which to undertake an offensive when we get a sufficiency of artillery,' Rawlinson argued, 'for observation is excellent and with plenty of guns and ammunition we ought to be able to avoid the heavy losses which the infantry have always suffered on previous occasions.' Rawlinson had great confidence in what could be achieved by the artillery. So certain was he that a thorough preliminary bombardment would crush the German defenders, that he convinced himself the infantry would have little difficulty crossing 'No-Man's Land', the name given to the open ground between the opposing lines of trenches, usually a few hundred yards across. This assumption had great repercussions, for Rawlinson consequently felt that it was unnecessary to provide his infantry subordinates with detailed advice concerning appropriate tactics.

Such apparent blind faith in the artillery certainly worried some of the men that were going to have to go over the top. Rifleman Percy Jones of the Queen's Westminster Rifles, 56th (London) Division, noted in his diary on 26 June 1916 that his divisional commander and his staff were 'busy telling us that we shall have practically no casualties because all the Germans will have been killed by our artillery barrage. This is nothing like the truth! The fact is that this attack is based entirely on supposition that there will be no Germans left alive to oppose us. On paper the plans are A1 [perfect], but if the Germans obstinately refuse to die and make way for us, our scheme will become impractical.' The assumptions that Rawlinson made about the artillery could have been challenged, but they were not. Indeed, when the commander in chief of the BEF, General Sir Douglas Haig, made revisions to the plan, he made the situ-ation worse, not better. Haig wanted a short 'hurricane' bombardment to create surprise, followed by an infantry advance towards deep first objectives which incorporated the entire German second line between the villages of Serre and Pozières and a line running from Pozières to the ridge northeast of Maricourt, south of the Albert–Bapaume road. Although Rawlinson managed to argue against the inclusion of the hurricane bombardment in the plan by saying that he thought that a protracted bombardment would have a greater effect upon German morale, he failed to put up much of a fight over the issue of objectives. As a result Rawlinson's plan was changed to incorporate Haig's wishes for deeper infantry objectives, but, crucially, no more guns were to be provided to attain them. This dilution of firepower was to have a catastrophic effect upon the opening day of the offensive.

Thus, owing to an ill-conceived compromise reached between Haig and Rawlinson, the plan that was eventually put into action was not the plan that either man wanted. The weaknesses of the plan, a fatal mixture of boldness and caution, were not immediately apparent. When the preliminary bombardment commenced on 24 June, its ferocity initially engendered a great spirit of optimism in many quarters on the British side, an optimism that up-beat intelligence reports did little to diminish.

Although some officers in the final few days of the artillery onslaught were sceptical about what it was actually achieving, their voices were drowned out by the roar of the guns.

The Attack

The British infantry marched several miles up to the front during the night of 30 June, each man carrying at least 66 pounds of equipment. By the time that they arrived at their positions in the early hours of 1 July, most were exhausted by a journey which ended in trenches made muddy by the rain that had fallen steadily during the previous few days. At 7.30 a.m., Zero Hour, the artillery barrage that had latterly been concentrating on the German front line moved forward and 66,000 men of the BEF's first wave left their trenches and started to move steadily across mostly open ground towards the German trenches. In the few minutes that followed, as the infantry entered a shell-holed wasteland, Germans began to climb out of their deep 'dugouts' (bunkers) to take up defensive positions in their damaged front line. At the same time, the German guns that had lain silent during the previous week began to open up, and a wall of explosives rained down on the British trenches and No-Man's Land. All along the front, the British infantry suffered such massive casualties that by noon few divisions north of the Albert–Bapaume Road were still moving forwards. The British attack stalled in a few hours. Although the fighting continued throughout the rest of the day, by nightfall it became clear that very few divisions had reached their first object-ives and many had even failed to reach the German trenches.

The failure of so many divisions even to cross No-Man's Land cannot be attributed to any single factor, but the failure of schemes designed to reduce German defensive effectiveness did little to help matters. British commanders were well aware that enemy strong points and deep bunkers could remain potential hazards even after their artillery had given them their full attention, and as a result plans were produced to tackle the problem. A diversionary attack was to be conducted by General Sir Edmund Allenby's Third Army with 46th Division and 56th Division around the village of Gommecourt, a short distance to the north of Fourth Army's line, in an attempt to take some German attention away. This attack went ahead on 1 July, but provided little in the way of diversion and cost 6,700

British casualties. Largely unsuccessful attempts to overcome the German defensive advantage were also tried, and included the explosion of a number of mines dug underneath enemy strong points just before Zero Hour. As one example, the mine that was detonated ten minutes before Zero Hour under the German trenches at Hawthorn Ridge Redoubt, just to the northwest of Beaumont Hamel, opposite the northern part of Fourth Army's line, did manage to destroy an important German strong point. But by announcing the imminence of the infantry assault, it also diminished what little surprise the British attack had left after a week-long preliminary bombardment.

The Failure of the Artillery Bombardment

Despite the planning and execution of the diversionary attack and the mines, the real power behind the reduction of German defensive positions was to be found in that preliminary bombardment in which Rawlinson put so much faith. The problem was that the bombardment failed to achieve what Rawlinson assumed it would achieve, the virtual annihilation of the German defences, because it was only half as intense as it needed to be. The preliminary bombardment was meant to carry out three crucial roles: to cut the barbed wire defending the German positions; to destroy the German trenches; and to severely reduce the German artillery. However, despite the one million shells devoted to the task, the wire was only successfully cut in front of XIII Corps at the southern end of the British line.

The difficulties that this caused the infantry approaching the German front line were tremendous. Banks of poorly cut wire lost attackers valuable time, and in many cases stalled the attack altogether. Either way, the failure to cut the wire completely meant that the crucial infantry momentum across No-Man's Land was lost and, as a consequence, the job of the German machine gunners was simplified. A good example of the sorts of losses sustained by the infantry who were confronted by uncut wire can be

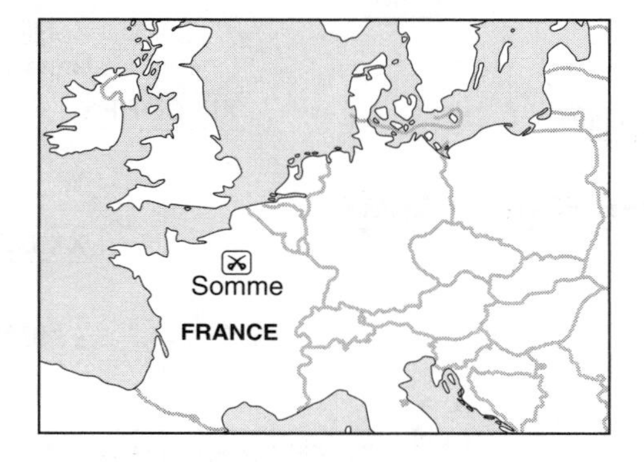

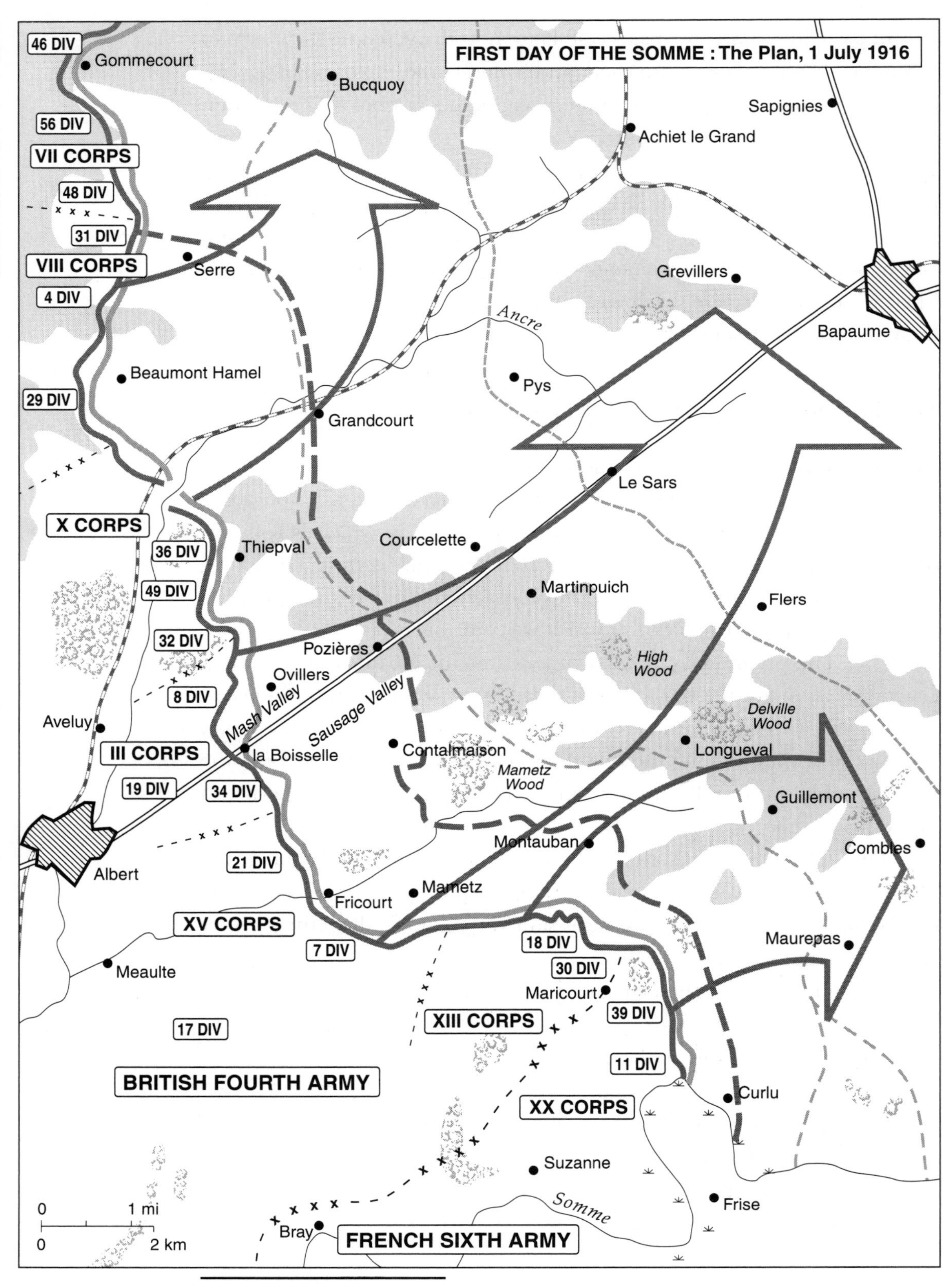
FIRST DAY OF THE SOMME : The Plan, 1 July 1916
46 DIV
Gommecourt
Bucquoy
Sapignies
56 DIV
Achiet le Grand
VII CORPS
48 DIV
31 DIV
VIII CORPS
Serre
4 DIV
Grevillers
Ancre
Bapaume
Beaumont Hamel
Pys
29 DIV
Grandcourt
Le Sars
X CORPS
Courcelette
36 DIV
Thiepval
49 DIV
Martinpuich
Flers
32 DIV
Pozières
High Wood
Ovillers
8 DIV
Mash Valley
Sausage Valley
Delville Wood
Aveluy
III CORPS
la Boisselle
Contalmaison
Longueval
Mametz Wood
19 DIV
34 DIV
Guillemont
Montauban
Combles
Albert
21 DIV
Fricourt
Mametz
XV CORPS
7 DIV
18 DIV
Maurepas
30 DIV
Meaulte
Maricourt
XIII CORPS
39 DIV
17 DIV
11 DIV
BRITISH FOURTH ARMY
Curlu
XX CORPS
Suzanne
Somme
Frise
0
1 mi
0
2 km
Bray
FRENCH SIXTH ARMY

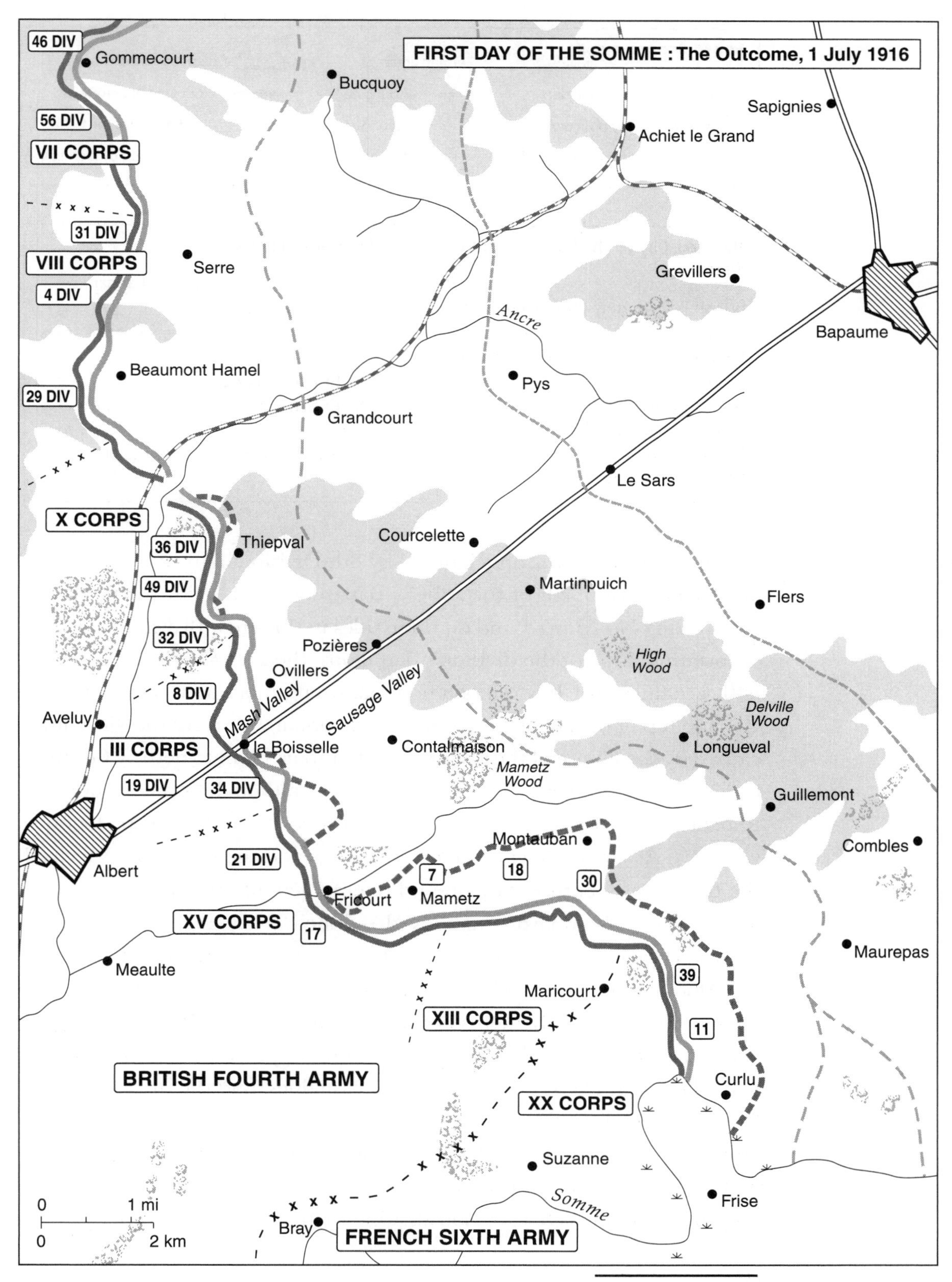
FIRST DAY OF THE SOMME : The Outcome, 1 July 1916
46 DIV
Gommecourt
Bucquoy
Sapignies
56 DIV
Achiet le Grand
VII CORPS
31 DIV
VIII CORPS
Serre
4 DIV
Grevillers
Ancre
Bapaume
Beaumont Hamel
Pys
29 DIV
Grandcourt
Le Sars
X CORPS
36 DIV
Thiepval
Courcelette
49 DIV
Martinpuich
Flers
32 DIV
Pozières
High Wood
Ovillers
8 DIV
Mash Valley
Sausage Valley
Delville Wood
Aveluy
III CORPS
la Boisselle
Contalmaison
Longueval
19 DIV
34 DIV
Mametz Wood
Guillemont
Montauban
Combles
21 DIV
7
18
30
Albert
Fricourt
Mametz
XV CORPS
17
Maurepas
Meaulte
39
Maricourt
XIII CORPS
11
BRITISH FOURTH ARMY
Curlu
XX CORPS
Suzanne
Frise
Somme
0
1 mi
0
2 km
Bray
FRENCH SIXTH ARMY

key to Somme maps	German lines	
- x x x - corps boundary	Front Line	planned Allied attacks
- x x x x - army boundary	second	Front Line at zero hour
high ground	third	Allied objective for 1 July
		positions held by nightfall

The First Day of the Somme, 1 July 1916

British Fourth Army under Lieutenant General Sir Henry Rawlinson
German Second Army under General Fritz von Below
British forces: 150,000
German forces: 30,000
British casualties: 57,470
German casualties: 8,000

Critical Moments

The British plan to seize the German second line defences
The preliminary British bombardment takes place
The British attack on 1 July
The British have partial success in the south
The British fail along the rest of the line

seen in the result of the attack put in by 8th Division in front of the village of Ovillers, just north of the Albert–Bapaume road. With the Germans overlooking No-Man's Land on three sides in this area, British losses were devastating. One of the division's battalions, the 2nd Middlesex, suffered 540 casualties on 1 July out of a notional strength of about 800 men.

On top of the problems caused by the uncut wire were the difficulties caused by the failure of the preliminary bombardment to destroy German trenches and bunkers. All along the front, although some sections of trenches were destroyed, substantial portions, including deep dugouts, were not. As a consequence, many Germans and their weapons survived the preliminary bombardment, and re-emerged into the sunlight as soon as the British bombardment moved off their front line at Zero Hour. German machine gunners on the high ground of Sausage Valley, south of the Albert–Bapaume road, survived the artillery onslaught and then had plenty of time to ready themselves for defence as 34th Division struggled over 500 yards of No-Man's Land and then stopped in front of the enemy wire. The 34th Division suffered 6,380 casualties, and the story was similar elsewhere along the front line. At Serre in the north of Fourth Army's attack, the German defenders used 74,000 rounds of ammunition to repel 31st (Pals) Division's attack and continued to fire until, at about midday, they had nothing left to shoot at. One of the division's battalions, the

Accrington Pals, suffered over 85% casualties on 1 July, or 585 men. One anonymous artilleryman watched the attack. 'Thousands went down that day', he wrote, 'I saw from my Post the first wave of troops scrambling out of their Trenches, in the early morning sunlight. I saw them advancing rapidly led by an officer, the officer reached the hillock holding his sword on high. Flashing it in the sunlight, he waved and sagged to the ground. His men undaunted swept up the mound to be mown down on reaching the skyline, like autumn corn before the cutter.'

The 'cutter' referred to was not just the machine gun, it was also the German artillery. In all 598 field guns and 246 heavier pieces survived the British counterbattery fire, and as a result they were able to bring down a hail of shells onto British positions. Because nothing could be done to penetrate this curtain of fire, the British troops that had made a lodgement in the German lines could neither be reinforced nor withdrawn, and most were eventually wiped out by counterattacks. This was a problem that 8th Division had to contend with as they attacked Mash Valley (next to Sausage Valley). Having attained a foothold in the German front line, the enemy shells that fell behind them cut them off from any potential support. This predicament was exacerbated by the fact that any subsequent movement forward was impossible, because of a creeping barrage fired in support which not only provided an incomplete shrapnel curtain owing to a lack of guns, but also lifted off the German front line too early and moved ahead of the infantry too fast.

The Few Successes

Despite these problems, some divisions north of the Albert–Bapaume road did manage to penetrate the German front line. Such success, however, was only ever on the front of a single division, and this in itself caused problems. The 36th (Ulster) Division managed to advance three-quarters of a mile into the German defences near Thiepval, but the failure of the divisions on either flank to make a similar advance meant that the Ulsters very quickly became isolated and highly vulnerable. With their lines of communication overstretched and their flanks wide open, the Ulster Division was gradually pushed back during the day, and by nightfall the troops were back in their own trenches again.

Many of the units that did manage some success on 1 July, the Ulsters included, employed innovative infantry tactics to aid their advance, rather than relying solely on the artillery. These tactics were used even though they had not been actively encouraged by Rawlinson, who thought that crossing No-Man's Land would not be difficult owing to sufficient artillery preparation. Some brigade and battalion commanders, perhaps worried about the assumptions made about the preliminary bombardment, ordered that troops move into No-Man's Land whilst the artillery concentrated on the German front line, so that they could then rush the enemy trenches when the barrage lifted. When 97th Brigade of 32nd Division used these tactics on 1 July, it succeeded in capturing Leipzig Redoubt near Thiepval; whilst its sister 96th Brigade used the tactics outlined in Fourth Army Tactical Notes and was cut down.

Sustained success, however, relied upon far more than just appropriate infantry tactics. The successes south of the Albert–Bapaume road of XV Corps, which captured all of its objectives on 1 July, and XIII Corps, which penetrated well into the German defences and held onto its position, came about as a result of a mixture of vital ingredients. In the southern sector of the Somme front where these achievements took place, the German defences had less depth than those further to the north. In the sectors of III Corps, X Corps, and VIII Corps north of the road, the German positions consisted of deep dugouts which were impervious to fire and constructed not only under the German front line, but also under several of the rearward lines. On the XV Corps and XIII Corps front, the German defences were concentrated in the front lines, which meant that in order for the Germans to make use of protection that the dugouts offered during the preliminary bombardment, they had to come well forward. Thus, with the British infantry starting out from advanced positions in No-Man's Land and the wire well cut, a high proportion of the German defenders were overrun in the first rush. Add to this the fact that on XV Corps and XIII Corps front there were fewer enemy guns, and the creeping barrage was more accurate and advanced at a slower pace than elsewhere, and the limited success south of the Albert–Bapaume road becomes easier to understand. It was a success that came partly from luck, and partly from disregarding Rawlinson's orders.

The Results of the Failure

British chances of success on the first day of the Somme depended upon how well the German defences had been destroyed, the speed of the infantry across No-Man's Land, and the weight of German artillery fire brought down upon the infantry as they advanced. As a result of this, although at the southern end of the British line much of the German front line was taken and held, north of the Albert–Bapaume road hardly a yard of German trench remained in British hands at the end of the day. Overall, the ground taken as a proportion of what was meant to been gained on the first day of the battle was tiny, despite costing the British 57,470 casualties including about 20,000 dead.

The disastrous first day of the Battle of the Somme was a calamity largely brought about by the decision reached by Haig and Rawlinson that, despite the number of guns available, deep objectives were to be taken. As a result, the preliminary bombardment was too diluted to successfully fulfil the vital roles given to it. Without proper battlefield preparation in the week running up to 1 July and useful fire support during the first day of the battle, asking the infantry to take even the German front line position would have been impossible.

Continued as a series of attacks through the summer and autumn, the Battle of the Somme eventually ended on 18 November, and quickly seared itself into the consciousness of the British nation. Although the battle can be viewed as an Allied attritional success, and as a battle from which the British Army learned a huge number of valuable lessons, it is the failure of its first day of the battle that will most readily be remembered.

LLOYD CLARK

Further Reading

Brown, M., *The Imperial War Museum Book of the Somme* (London, 1996)

Middlebrook, M., *The First Day on the Somme 1 July 1916* (London, 1971)

Prior, R. and Wilson, T., *Command on the Western Front* (London, 1992)

Turner, W., *Pals – The 11th (Service) Battalion, East Lancashire Regiment* (London, n.d.)

The Battle of the Bulge

16 December 1944–28 January 1945

Progressive Unreality

'All Hitler wants me to do is to cross a river, capture Brussels, and then go on to take Antwerp! And this in the worst time of the year through the Ardennes, where the snow is waist deep and there isn't room to deploy four tanks abreast let alone four armoured divisions. Where it doesn't get light until eight and it's dark again at four. And with re-formed divisions made up chiefly of kids and sick old men. And at Christmas!' Colonel General Sepp Dietrich

For generals to base their planning on one major factor, which turns out to be wrong, is surprising, but eventually understandable. More rarely, a battle plan is made which depends for its success on a series of assumptions, each one more unlikely than the next. Military history is full of stirring tales of adventurous plans for attacks that succeeded against fearful odds, despite predictions that they would fail. This is the story of a desperate attack, launched in the last months of World War II in Europe, which the senior commanders on *both* sides believed would inevitably fail, and which did.

The Origins of the Battle

The story of the Battle of the Bulge starts on 20 July 1944, during the disastrous German defeats by the western Allies in Normandy and the Soviet forces in central Europe. Officers of the German Army, convinced that Germany would inevitably lose the war, tried and failed to assassinate Adolf Hitler with a bomb at his headquarters, the 'Wolf's Lair' near Rastenburg in eastern Germany. A few days later, German attempts to hold their ground in Normandy finally collapsed. During August and September, the retreating Germans were harassed and chased by the Allies through France, Belgium, and the southern Netherlands, finally managing to stabilize their line on the western border of Germany itself, protected by the fortifications of the *Westwall*, known to the Allies as the 'Siegfried Line'. But all the while, Hitler was plotting a great counterattack that would bring him victory. He called for even greater efforts from Germany to build the tanks, guns, and aircraft for this last attack. At the same time, increasingly he placed his trust not in his Army commanders but in the Waffen-SS, the Nazi Party's own armed forces, which had grown to massive size during the war.

The Allied overall commander was US General Dwight D Eisenhower, whose SHAEF headquarters (Supreme Headquarters Allied Expeditionary Forces) was at Versailles near Paris. Under Eisenhower were three strong army groups: in the north the Anglo-Canadian 21st Army Group under the British Field Marshal Sir Bernard Montgomery, in the south the French and American 6th Army Group under US Lieutenant General Jacob Devers (which took no part in the Battle of the Bulge), and between the two the entirely American 12th Army Group under US Lieutenant General Omar Bradley. Under Bradley's command were the weak Ninth Army next to the British First Army under Lieutenant General Courtney Hodges in the centre, and the powerful Third Army under Lieutenant General George Patton in the south.

On the German side of the line, opposing Eisenhower as theatre commander was the veteran Field Marshal Gerd von Rundstedt at *OB West* (*Oberkommando West*, or German High Command West), with four severely weakened army groups, of which the strongest, Army Group B in the centre under Field Marshal Walter Model, was intended as Hitler's instrument of victory.

The Allied Plans

Strictly, the Allies had no plans for the forthcoming battle, which caught them completely by surprise. Although the Allies largely stopped offensive operations during November and December, some attacks were made to breach the Westwall defences, in particular by Patton near Saarbrücken and Hodges near Aachen, in order to provide jumping-off positions for major offensives in the spring, when Montgomery and Patton would each make a big pincer movement towards the industrial cities of the Ruhr. To this end, First Army was deployed with two of its three corps concentrated around Aachen, and with the third, VIII Corps under Major General Troy Middleton, stretched out through the very broken, wooded and hilly country of the Ardennes forest in Luxembourg and southern Belgium. This was the least promising terrain of all for an armoured attack, although it had been crossed by German armour in 1940. Middleton's orders were to 'roll with the punch' in the unlikely event of a German attack, but otherwise VIII Corps sector was seen as a rest and training area. Germany was surrounded by enemies from all sides with no possibility of help. The Allies had superiority on the ground and virtually complete control of the sea and the air. Simply, the idea of Germany attacking did not make sense to the Allies; there was nothing that such an attack could achieve. Growing intelligence reports of a German attack were dismissed as impossible. Although the Allied high command was later much criticized for failing to believe in the German attack, they were in fact quite right to see it as utterly unrealistic. It is a common military saying that you can always get surprise by doing something so stupid that it cannot work.

The German Plans

The view from the Wolf's Lair was quite different. Hitler accepted that the war was lost, but only as long as the 'Big Three' Allies – the Soviet Union, the United States, and the British Empire – held together. If Britain and the United States could be brought to a compromise peace, then Germany might still be strong enough to defeat the Soviet Union alone. This political fantasy became the first stage of the progressive unreality that led to the plan for the Battle of the Bulge. Model's Army Group B was given Germany's last armoured reserves: Sixth Panzer Army under SS Colonel General Josef ('Sepp') Dietrich and Fifth Panzer Army under General Hasso-Eccard von Manteuffel. These two armies were to attack in bad weather to minimize the effects of Allied air power, punching through the weakest part of the Allied line in the Ardennes, the front held by Middleton's VIII Corps. With their southern flank protected by General Eric Brandenburger's Seventh Army, they would capture crossings over the River Meuse within two days. Within another two days Dietrich's formidable spearhead of four Waffen-SS Panzer (armoured) divisions would drive on to capture the main Allied supply port of Antwerp, with Manteuffel in support, so encircling and trapping 21st Army Group and much of 12th Army Group with it. The British and Canadians, having no way to replace their destroyed formations, would be effectively out of the war; while the weakened Americans, heavily committed against the Japanese in the Pacific, could be expected to negotiate peace rather than allow a victorious Soviet Union to dominate Europe.

This plan was based on an almost contemptuous assessment of the Anglo-American alliance and its armed forces: that its commanders would be slow to react, distrustful, and uncooperative with each other, and that its troops would run away or collapse when attacked or surrounded. It also envisaged rates of advance through bad terrain in worse weather that were utterly unrealistic for German forces that were scraping the very bottom of the barrel. From Rundstedt downwards, the German commanders charged with carrying out this plan rejected it as unworkable. Even Dietrich, a former commander of Hitler's SS bodyguard, poured scorn upon it. Model offered instead the 'Small Solution', a pincer move by Army Group B from north and south to destroy most of First Army east of the Meuse, so weakening and delaying the Allied spring offensive.

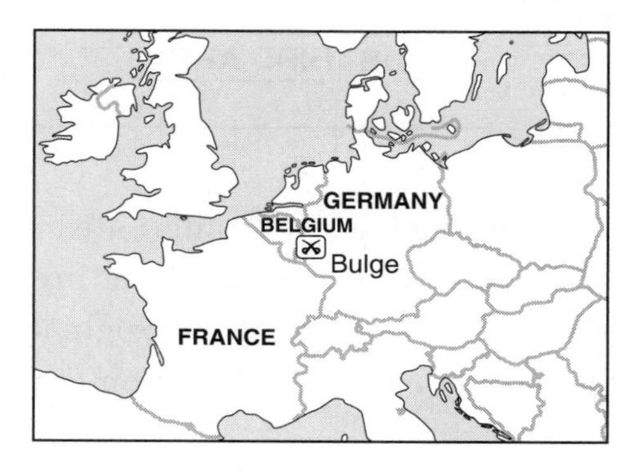

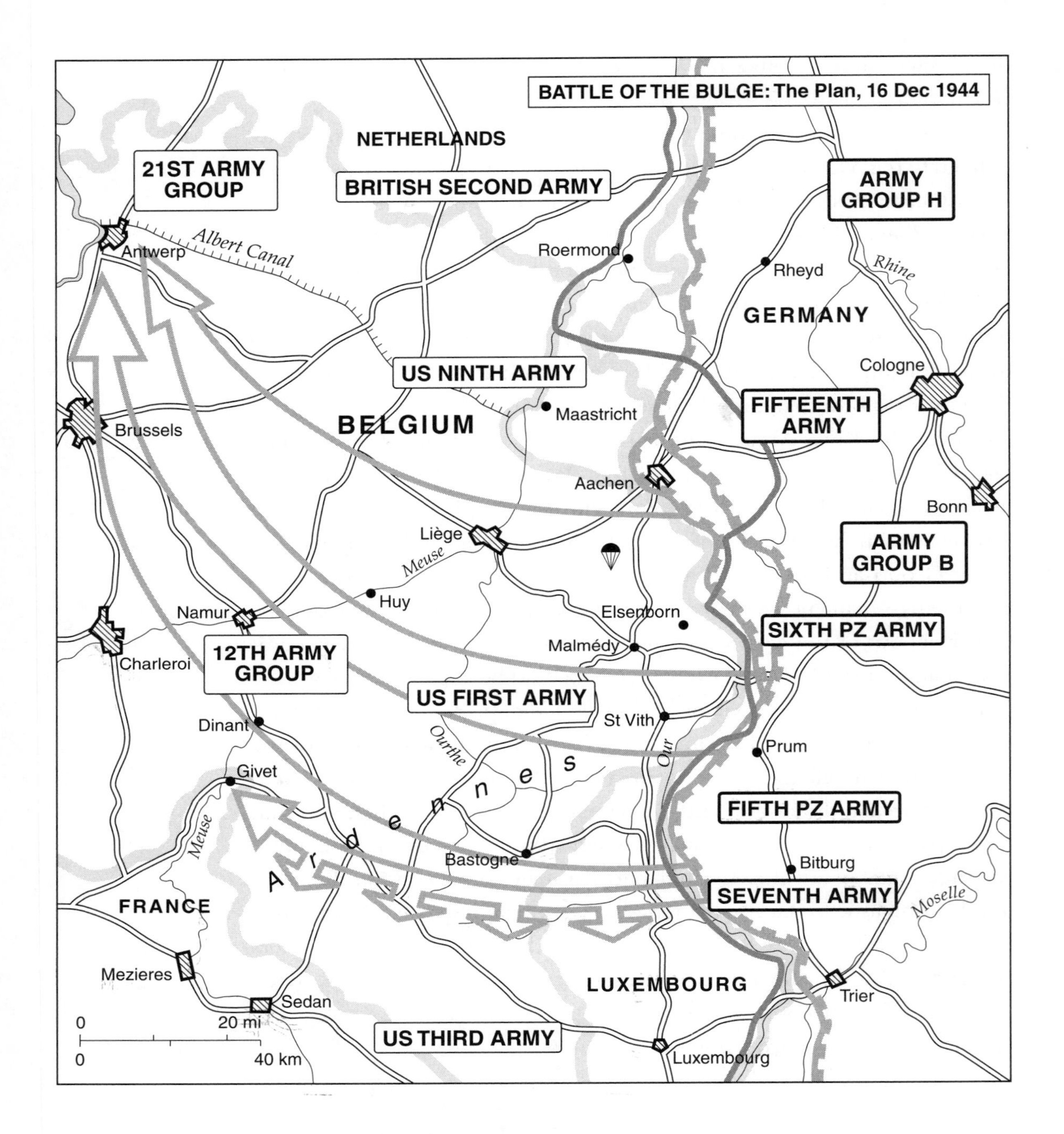

Although Model's alternative plan might have stood some chance of working, it would not have prevented the Allies winning the war. Hitler's insistence that the original plan was carried out was based on the utterly ruthless calculation that whatever the odds against its suc-

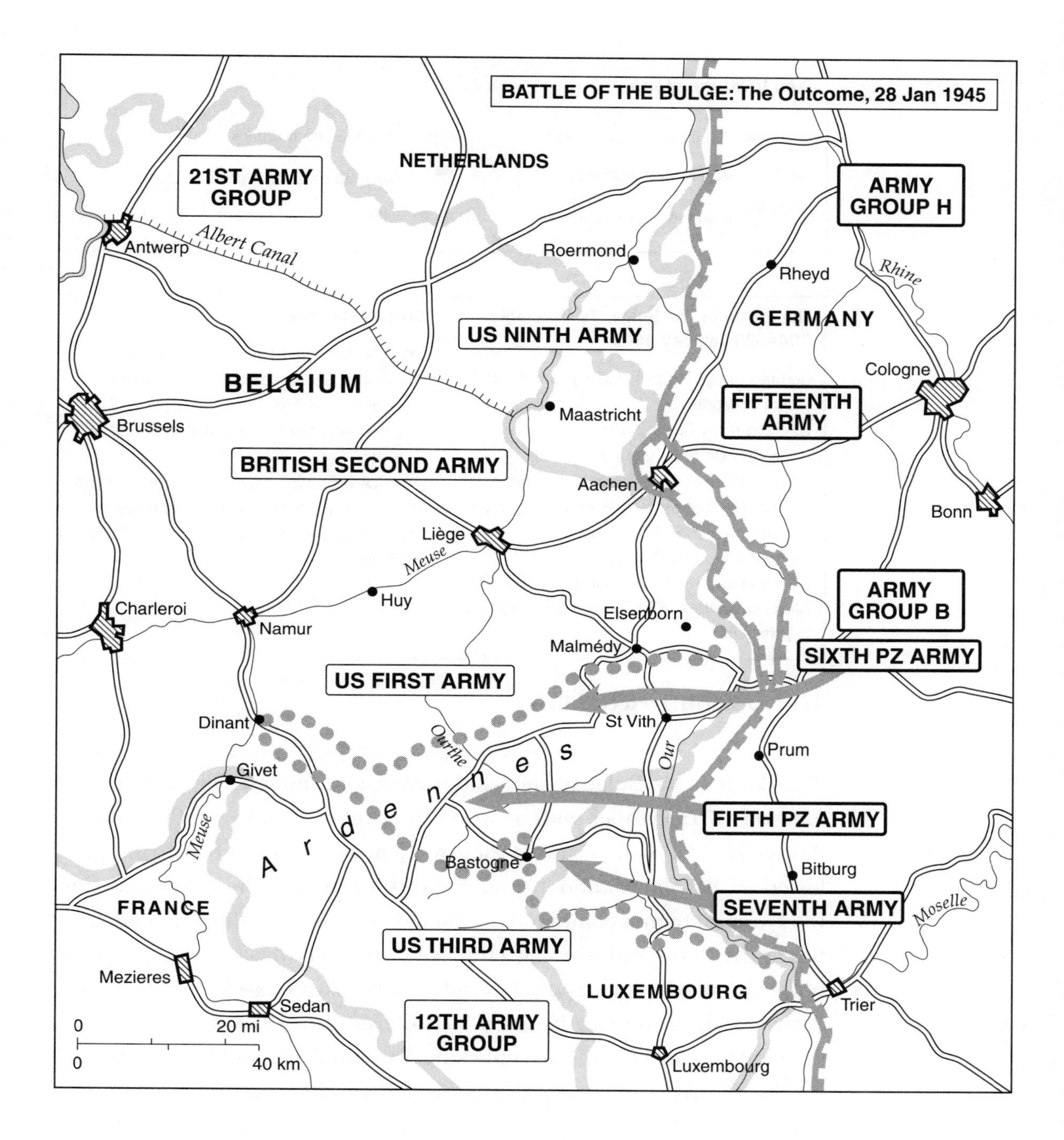

cess (and no matter how many Germans died in the attempt) it remained the one chance for victory. Even some of the troops shared this view, particularly in the Waffen-SS, preferring to seek death in battle rather than live to see defeat.

key to Bulge maps

Front Line 15 Dec 1944	actual German attacks
Westwall (Siegfried Line)	German parachute drop 16 Dec 1944
planned German attacks	limit of German penetration 26 Dec 1944

The Battle of the Bulge, 16 December 1944–28 January 1945

German Army Group B under Field Marshal Walter Model
US 12th Army Group under Lieutenant General Omar N Bradley
Anglo-Canadian 21st Army Group under Field Marshal Sir Bernard L Montgomery
German forces: 500,000 troops
Allied forces: 655,000 troops
German casualties: approximately 100,000
Allied casualties: 82,400 (including 15,000 prisoners)

Critical Moments

The Germans plan to advance on Antwerp
The initial German attack stalls on the Elsenborn Ridge
The Germans fail to take Bastogne and St Vith
American reserves move down from Aachen
Patton's Third Army relieves Bastogne
The Germans are halted short of the River Meuse

The German Attack

The Battle of the Bulge opened in darkness, fog, and freezing snow at 5.30 a.m. on 16 December with a two hour bombardment by 2,000 guns of Army Group B which caught the Americans completely by surprise, followed by an infantry and armoured advance. At first there was little that the German higher commanders could do to influence the outcome, having few reserves and needing Hitler's permission to use them. Almost at once, the plan went wrong. Instead of a breakthrough by Sixth Panzer Army on the boundary between VIII Corps and V Corps under Major General Leonard Gerow, the attacking armoured forces of I SS Corps and II SS Corps encountered unexpectedly stiff resistance from the Americans, who after two days of hard fighting had been pushed back only about five miles to the secure high ground of the Elsenborn Ridge, against which no further progress could be made.

A German attempt to assist the advance in the north by dropping an improvised force of paratroopers on the night of 16–17 December was badly scattered by crosswinds and achieved nothing. But the use of a special unit of German troops dressed and equipped as Americans did achieve its objective of creating confusion in the Allied rear areas, and for a time even Eisenhower was forbidden to leave his Versailles headquarters in case the Germans tried to assassinate him.

Instead of the decisive breakthrough by Sixth Panzer Army as far as the Meuse, its one success was the remarkable advance of 'Kampfgruppe Peiper', the leading battlegroup of 1st SS Panzer Division, which consisted of about 4,000 men and 100 tanks under the 29-year-old Lieutenant Colonel Joachim ('Jochen') Peiper, and which managed to force its way through the American defences. By nightfall on 19 December Kampfgruppe Peiper had driven forward unsupported about 20 miles behind the original American front line, but was still only half way to the Meuse crossings.

While Sixth Panzer Army was stalled, the greatest German success came where it had not been expected, from Fifth Panzer Army in the centre, where LXVI Corps brought off a spectacular double encirclement which virtually eliminated the novice American 106th ('Golden Lions') Division, while the armour of XLVII Panzer Corps and LVIII Panzer Corps drove for the critical road junction of Bastogne. But although caught out by the German attack, the Americans once more fought back in a way that their enemies had not expected, with small groups of men fiercely contesting the woods, hills, and river crossings of the Ardennes, aided by Allied airforces that were still trying to fly through the fog and overcast. As German tanks broke past their first defences, the Americans improvised combat teams from cooks, orderlies, and headquarters clerks. Across the front, hundreds of little fire-fights delayed the German advance and progressively wrecked its timetable. When the leading tanks of XLVII Panzer Corps reached Bastogne on the morning of 19 December it was already too late.

At Allied high command, no one could at first believe that this was a major German offensive. But late on 16 December Bradley, who was visiting SHAEF headquarters at Versailles, began ordering reinforcements to help Middleton, obtaining more from Eisenhower's strategic reserve. By the night of 18 December retreating forces from VIII Corps gathering at Bastogne had been met by more armoured troops and by the elite 101st ('Screaming Eagles') Airborne Division, which had moved in by road. The 101st Airborne's senior officer present, Brigadier General Anthony

McAuliffe, took command of the Bastogne garrison, which defied German attempts to storm its defensive perimeter next day. To maintain momentum the German armour by-passed Bastogne, leaving the infantry to capture it. Meanwhile, 7th Armoured Division moving down from the north also arrived on 19 December at St Vith, the other critical road junction on the battlefield, forming the basis for its continued defence.

Rather than being across the Meuse and advancing on Antwerp, the Germans were being held in the north and delayed in the centre and south, still well short of their first objectives. What was more, there was little that their commanders could do except hope that the Americans would somehow break and allow them to drive through. The utter un-reality of the German plan for the battle was becoming obvious to both sides.

The Allied Counterattack

On the morning of 19 December Eisenhower called a conference at Bradley's headquarters in Verdun for his senior commanders. In the middle of fighting the battle, Hodges obviously could not attend. Instead, he was progressively redeploying his powerful VII Corps under Major General J Lawton Collins from Aachen to seal off the northern part of the battlefield and prevent any reinforcement of Kampfgruppe Peiper. Eisenhower began the meeting by assessing the German attack as bound to fail, and looked for ways to destroy it. First he ordered Devers to put Sixth Army Group on the defensive in case of a further German attack to the south. (This actually took place in Alsace on 31 December, achieving nothing.) The Verdun conference showed just how wrong the Germans had been about the Allied commanders' willingness to cooperate with each other. Patton volunteered to halt Third Army's attacks against the Westwall and to swing half his forces, III Corps, and XII Corps, through ninety degrees to drive into the German southern flank held by Brandenburger. The orders to start this astonishing move were already issued, and Patton agreed to commence his attack within three days.

Later Montgomery, who had not been at the conference but had followed his normal practice of sending his chief of staff, called Eisenhower's headquarters at Versailles with his own offer of help. The battle was dividing into two separate operations, one in the south to relieve Bastogne and the other in the north to prevent the Germans making further headway.

To reflect this, Eisenhower agreed to Montgomery's suggestion that the boundary between 21st Army Group and 12th Army Group should be moved south to run through the centre of the battlefield, from Givet on the French–Belgian border to Prüm in Germany. This placed Ninth Army and most of First Army under Montgomery, and allowed him to send British XXX Corps to defend the Meuse crossings against the increasingly unlikely event of the Germans breaking through. This in turn allowed Hodges to focus entirely on directing his own forces to defeat the Germans.

These command decisions, which Hitler and his planners had assumed would take several days of debate in Washington and London, were taken by Eisenhower personally within a day of a clear picture of the German attack emerging. This quick decision-making, together with the strong resistance of the American troops in battle, destroyed any hope of a German victory. On 20 December Hitler at last conceded that Dietrich was not going to break through in the north and allowed II SS Corps to be transferred to Fifth Panzer Army. After a hard-fought battle the Americans finally pulled back from St Vith on 23 December. It was much too late for the Germans: on the same day the trapped Kampfgruppe Peiper finally dispersed or surrendered. A day earlier, Patton's attack to relieve Bastogne had started. McAuliffe, summoned to surrender by the Germans, had replied with a note containing one of the most famous messages in warfare: 'To the German commander – NUTS! – the American commander.' What was worse for the Germans, the skies above the battlefield were clearing, and the Allied airforces could come fully into play.

The high point of the German advance was reached on 26 December, producing the 'Bulge' in the American line which gave the battle its name. In the north, Sixth Panzer Army was still locked in combat with the Americans on the Elsenborn Ridge. In the south, Seventh Army was trying to halt Patton's offensive. Between the two, Manteuffel's Fifth Panzer Army lay spread out through the hills and valleys of the Ardennes. Supply and transport problems meant that German tanks were running dry on the battlefield for lack of fuel. Although a last attack at Bastogne on 25 December – Christmas Day – almost drove to the town centre, it was finally repulsed. McAuliffe was already getting supplies from the air, and on the afternoon of 26 December the leading tanks of 4th Armoured Division broke through from the south to end the siege. Just one German column reached a crossing over the Meuse, at Dinant on Christmas Day, only to find it guarded by British tanks. On the same day the last attempt by XLVII Panzer Corps to make forward progress towards the Meuse was smashed by an American counterattack led by 2nd ('Hell on Wheels') Armoured Division, still well short of the river.

Reality Sinks In

The Germans had now lost the initiative, and despite repeated attacks against Bastogne they could not recover it. By 30 December their last attempts had failed against a much strengthened American line. On 1 January 1945, in what the Americans called 'the hangover raid', the Germans threw their last remaining reserves of aircraft into the battle with a surprise dawn attack on 27 Allied airfields. Of 1,035 German aircraft taking part over 300 were shot down, compared with 156 Allied aircraft.

Having been surprised by the Germans once, the Allied commanders chose caution for their own counterattack, despite their greatly superior strength. By 3 January a general attack was in progress from Montgomery's forces in the north and Bradley's in the south, between them squeezing out the Bulge, while Allied aircraft struck at the retreating Germans. It was a slow but relentless drive, taking a month to recapture the ground that the Germans had seized in ten days. On 28 January, back on their original line, the Americans pronounced the battle officially over. From the German perspective it had never stood any chance of success. Hitler and some of the Germans had talked themselves into believing that the Americans were bad troops and the Allied generals were bad commanders. The Battle of the Bulge, planned on that basis, had shown them to be utterly wrong.

STEPHEN BADSEY

Further Reading

Dupuy, T.N., *Hitler's Last Gamble* (New York, 1994)

Eisenhower, D.D., *Crusade in Europe* (New York, 1948)

Liddell Hart, B.H., *The Other Side of the Hill* (London, 1948)

MacDonald, C.B., *The Battle of the Bulge* (New York, 1984)

Underestimating the Enemy

'Know the enemy and know yourself, and in a hundred battles you will never be defeated; when you are ignorant of the enemy but know yourself, your chances of winning or losing are equal; if you are ignorant of both your enemy and yourself, you are certain to be defeated.' Sun Tzu, *The Art of War*

How many battles in history have been lost because someone made a plan which failed to take into account that the enemy might be better than they were thought to be; that they might have a genius for a commander, or move faster than they were supposed to do, or turn out to have a military capacity and flair that was not suspected before the battle began? A good battle plan must always take the enemy into account. While every commander must have confidence in himself and in his army, overconfidence leads all too often to defeat.

Austerlitz

2 December 1805

Underestimating Napoleon!

'They seemed to forget that they were dealing with the greatest commander in the world... that even his apparently unconsidered actions were the direct reflections of some very deep thinking.' Brigadier General Thiebault of IV Corps

How does one plan to defeat the greatest military genius of the age, at the very height of his powers, and with an army probably unrivalled in Europe for a century? Still towering over the elegant Place d' Austerlitz, in the heart of Paris, is the memorial column made from the bronze of 180 Austrian and Russian cannon captured during Napoleon's victory near the town in Moravia whose name both the square and the column bear. It is a memorial not only to Napoleon and his *Grande Armée*, but to the defeat of his Austro-Russian opponents, whose planning for the battle revealed an almost breathtaking failure to realize the nature of what they were facing. In a single blow at Austerlitz, the Third Coalition against Napoleon (composed chiefly of Great Britain, Austria, and Russia) was destroyed, and the young Imperial system that they had threatened to overwhelm was secured. Austerlitz was Napoleon's greatest victory, and although there would be others in later years, none would be as spectacular or as comprehensive.

Napoleon Moves First

'My mind is made up ... by September 17, I shall be with 200,000 men in Germany.' With those words to his Foreign Minister Prince Tallyrand on 25 August 1805, Napoleon Bonaparte discarded his abortive plan for a 'descent upon England'. Over the next few days La Grande Armée abandoned its encampments around the port of Boulogne, and marched towards France's eastern border. This hasty and dramatic change in Napoleon's plans had been forced upon him by the patient and skilful diplomacy of the British Prime Minister, William Pitt ('the Younger'). Exploiting European fears of the growing dominance of France, Pitt had managed to seduce an initially suspicious Russia and a hesitant Austria into an alliance with Britain. The Third Coalition intended to redress its signatories' various grievances against France, and finally to cut the upstart Corsican down to size.

Napoleon did not wait for the Allies to attack, but instead decided to launch a pre-emptive blow before their forces were ready. In a lightning advance to the Danube in just 26 days, with little fighting, the French encircled General Karl Mack's army of 60,000 troops in the vicinity of Ulm and forced them to surrender. With the road to Vienna now open, Napoleon moved swiftly to catch two Russian armies advancing from the east before they could withdraw. But despite a vigorous pursuit, which saw Vienna taken on 12 November, the Russians successfully evaded La Grande Armée. On 23 November, with the weather worsening and his army exhausted after eight weeks of continual rapid marching, Napoleon was forced to halt the advance near Brunn, 60 miles north of the Austrian capital.

Napoleon now found himself in a perilous situation. Losses in the course of the arduous autumn marches, desertion, and the need to detach forces to cover the army's flanks and rear, had drastically diminished the strength of La Grande Armée. Scattered across Moravia, the French had only their I Corps, IV Corps, and V Corps, a division of Grenadiers, Marshal Joachim Murat's cavalry reserve, and the Imperial Guard, in all some 60,000 troops. The onset of winter and food shortages would continue to weaken French strength. In contrast, the concentration of two Russian armies, led by Marshal Mikhail Kutuzov and Lieutenant General Friedrich Buxhowden, as well as Prince Johann of Liechtenstein's

Austrian force, gave the Allies a combined strength of 80,000 troops. The future arrival of reinforcements from Italy and Russia would give the Allies an overwhelming numerical superiority over the French. It was clear that the initiative was on the point of changing sides, and that unless the Austro-Russian Army could be defeated in the near future, Napoleon faced two stark alternatives: to withdraw or be defeated.

The Tsar Takes Command

Initially the senior Allied commanders – Kutuzov, Prince Peter Bagration, Mikhail Miloradavich, Dmitri Dokhturov, and Andrault Langeron – decided to take no risks and let time and weather take their toll on the French. In light of their recent reverses they had developed a strong respect for their enemy's capabilities, and especially for Napoleon's generalship. This cautious attitude was replaced by a rash and impetuous approach after the arrival of Tsar Alexander I at Kutuzov's headquarters at Olmutz on 22 November. The Tsar and his personal companions were full of self-confidence. They openly despised the Austrians, believing that only Russian troops were worthy opponents of the French. Russian generals who advocated caution were treated with contempt and their advice ignored. When Kutuzov expressed his views about the future movements of the army he was curtly informed 'that is none of your business!' Allied strategy was now dictated by the Tsar and his young friends, who confidently believed that in one stroke they could beat a man who as yet had no rival, let alone an equal, on the battlefield.

The change in attitude at Allied headquarters was encouraged further by Napoleon's attempts to lure the Austro-Russian Army into a premature attack by feigning weakness. On several occasions his *aide-de-camp*, Brigadier General A J M R Savary, visited the Austro-Russian headquarters to discuss the possibility of an armistice. Napoleon hoped that these false requests would convince Allied commanders that the French were vulnerable and desperate to avoid battle. Napoleon reinforced this impression of weakness by ordering his troops to withdraw in the face of the Allies when they began their advance on 27 November. The order to abandon the strong defensive positions on the dominating Pratzen Heights without a fight on 1 December was the final step in persuading the Allies that La Grande Armée was near to collapse.

Napoleon's Plan

As the Austro-Russian Army advanced towards Austerlitz, the French swiftly concentrated their dispersed forces. On 1 December the main body of the army was deployed west of the Pratzen Heights around the Zuran Hills which concealed it from the Allies. The French left flank rested on the strong position around the fortified Santon Hill, whilst to the south the right flank covered the line of the Goldbach stream from Puntowitz in the centre to Tellnitz in the south. Despite the best efforts of the French commanders to collect as many troops as possible, La Grande Armée could muster only 60,000 troops to oppose an Allied force of 80,000. Although Marshal Louis Davout's III Corps, marching from Vienna, could be expected to join the army during the morning of 2 December, this would only add a meagre 5,000 troops.

Napoleon was depending upon the Allies making a serious error in the disposition of their forces and providing him with the opportunity to deliver a crushing blow. His plans had been taking shape for some time before 2 December. When French forces had first occupied the Pratzen Heights and the surrounding area on 20 November, Napoleon had told his commanders, 'Take good note of this high ground. You'll be fighting here before two months are out.' The concentration of Allied forces towards the southern end of the Pratzen position made it obvious that their main attack would fall along the Goldbach, between Tellnitz and the Kobelnitz pond, with the intention of cutting the French communications to Vienna. Napoleon realized that in the course of this movement the Allies would leave the Pratzen Heights unoccupied, and expose their right flank and rear to French forces farther north.

To encourage the Allied advance, only a light screen of troops defended the lower Goldbach. Davout's forces, expected to arrive later in the morning, were to deploy on the plain behind the stream near Turas, and expel any enemy forces that had advanced so far north. When the Austro-Russian forces were fully committed to crossing the Goldbach, the main body of the French army was to wheel south over the Pratzen Heights and attack their exposed right and rear. Marshal Nicholas Soult's IV Corps would spearhead the advance from its positions near Puntowitz, and seize as much of the Pratzen Heights as pos-sible. Once the rest of the army – Marshal Jean Baptiste Bernadotte's I Corps, Murat's cavalry, and

Marshal Jean Lannes' V Corps – had joined Soult from their positions further north around the Zuran Hills, 53,000 French troops would be unleashed in a devastating attack from the Heights against the congested Allied columns in the valley below. The Imperial Guard commanded by Marshal Jean Bessieres, together with General Nicholas Oudinot's Grenadier division, would remain in reserve. In the early hours of the morning of 2 December, General Claude Legrand's 3rd Division was detached from IV Corps to reinforce the troops defending the line of the Goldbach.

The Allied Plan

To the east of the French campfires, Allied forces along the Pratzen Heights were preparing for the coming battle. In the early hours of 2 December the Austrian General Franz Weyrother outlined the plan of attack to his fellow commanders. Noticing the weakness of French forces along the Goldbach, the main Austro-Russian effort would take the form of a wide left hook across the stream between the villages of Tellnitz and Puntowitz. In the area between Tellnitz and the Kobelnitz pond, Lieutenant General Michael Kienmayer's advance guard was to secure the left bank of the Goldbach, and cover the deployment of Buxhowden's forces. The latter consisted of approximately 40,000 troops divided into three columns, commanded by Lieutenant Generals Dokhturov, Langeron, and Ivan Prebyshevsky. A fourth column of 23,000 troops, led by Lieutenant Generals Miloradavich and Johann Kollowrath, was to cross the Goldbach north of the Kobelnitz pond.

Once these preliminary movements were complete the main attack would commence. The Allied columns would wheel north and drive the French right flank into the main body of La Grande Armée, which in the ensuing disorder would be destroyed. A separate column of 13,000 troops led by Prince Bagration was to block the Brunn–Olmutz road, to prevent the French escaping eastward. The gap between Bagration and the

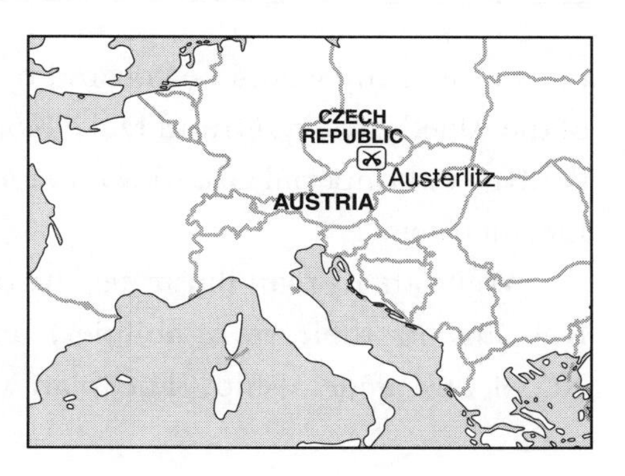

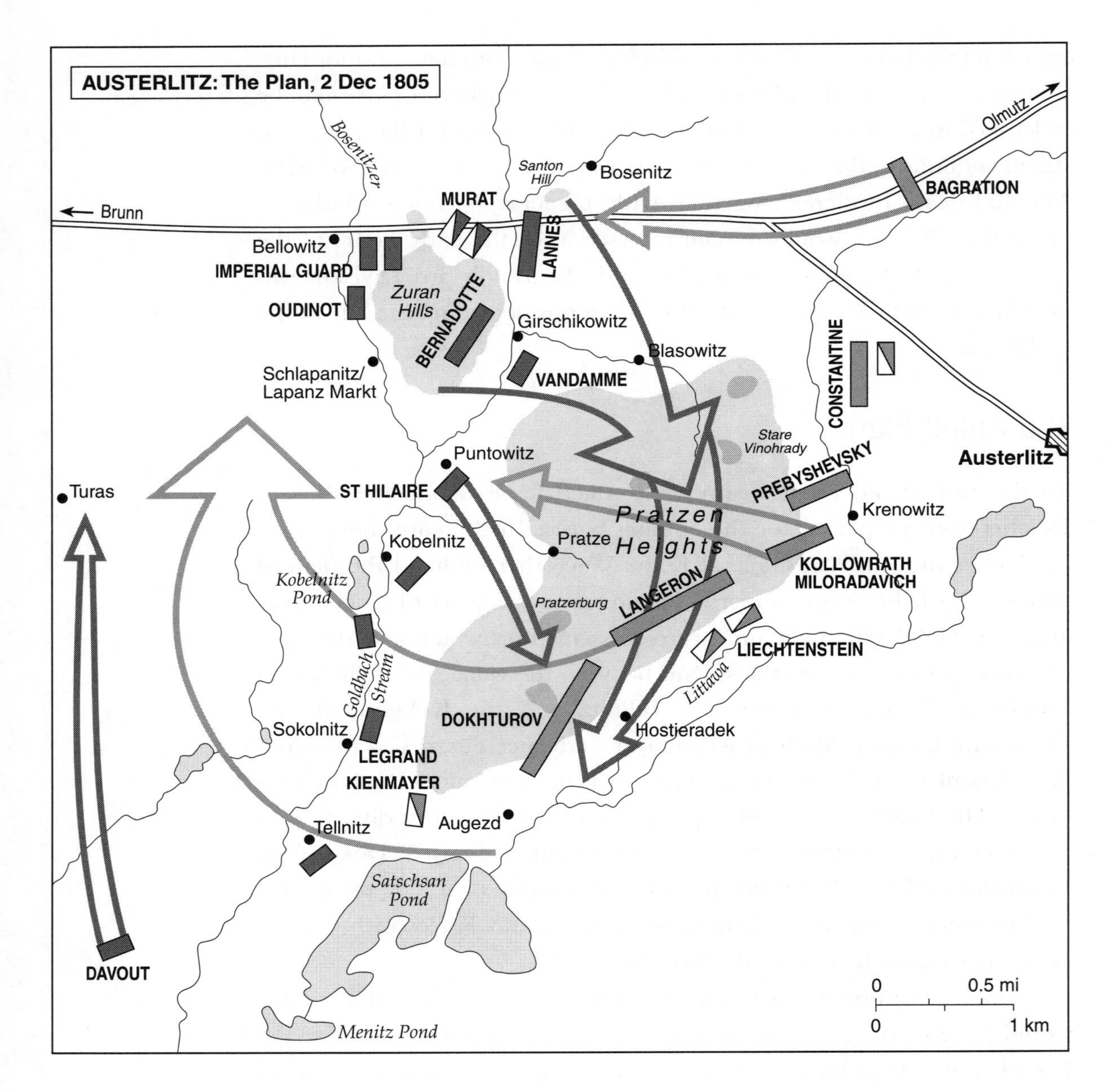

rest of the army would be covered by Prince Liechtenstein with the bulk of the Allied cavalry. Grand Duke Konstantin Pavlovich Constantine with the Russian Imperial Guard was to form a small reserve for Bagration and Liechtenstein.

Weyrother's plan illustrates the overconfidence of the Austrians and Russians in their own abilities, and their fatal misappreciation of Napoleon's generalship. The plan was too ambitious for the Austro-

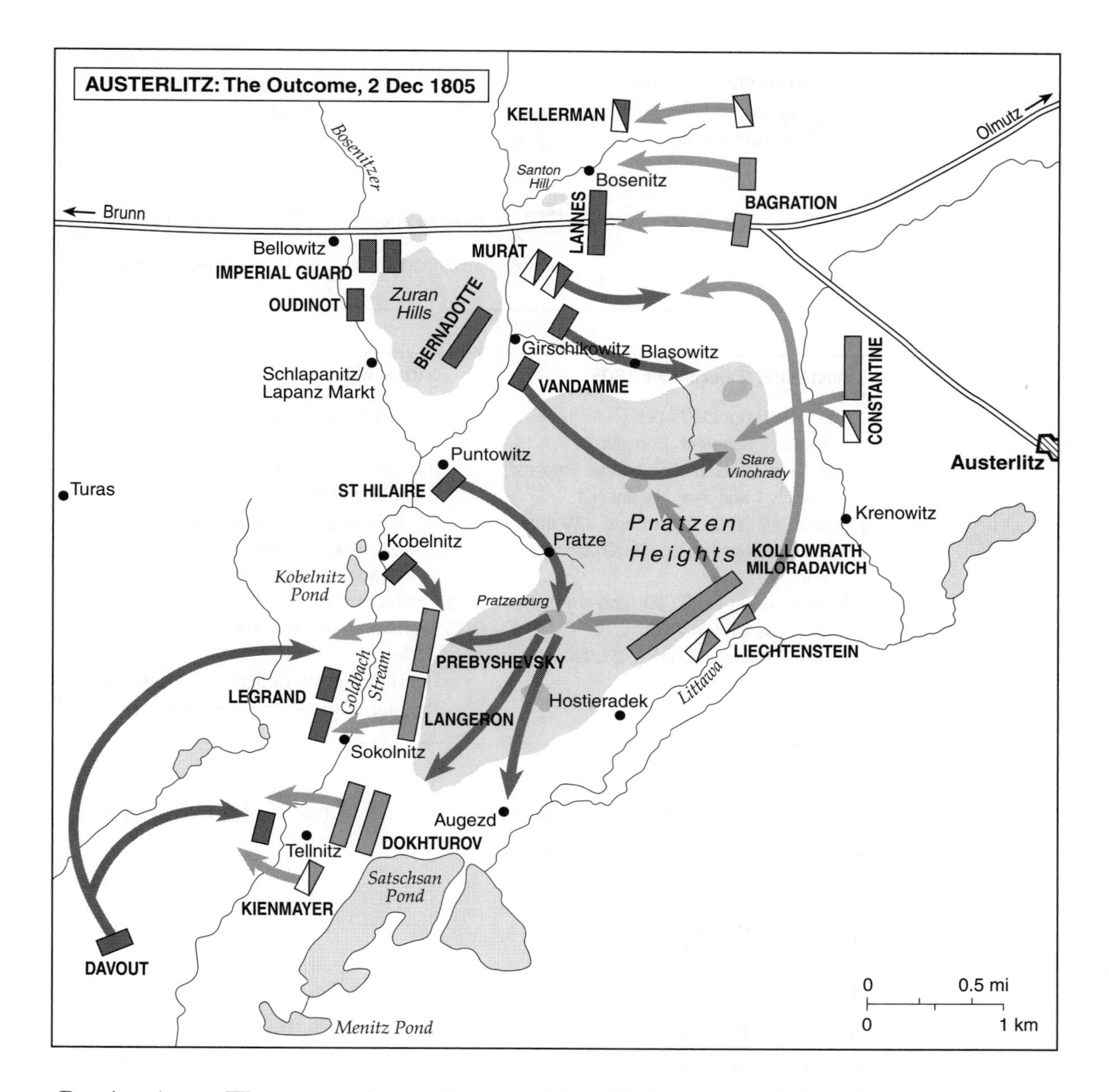

Russian Army. The poor training of many of the Allied troops, and the ad hoc composition of the four Allied attacking columns, meant that there was little cohesion between the regiments comprising them. The plan also presupposed a level of coordination between the columns that was not realistically possible. A second flaw was that there was no allowance for a French defence of the Goldbach, or even the possibility of the French mounting an attack. The only concern of Allied command-ers during the

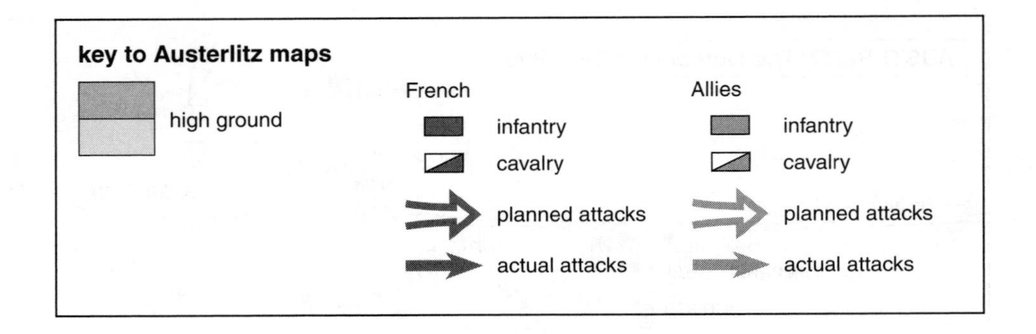

Austerlitz, 2 December 1805

The French Grande Armée under the Emperor Napoleon Bonaparte

The Austro-Russian Army under Emperor Francis I and Tsar Alexander I

French forces: 65,000 troops, 139 guns

Austro-Russian forces: 80,000 troops, 278 guns

French casualties: 8,000 (two-thirds from IV Corps alone)

Austro-Russian casualties: 27,000

Critical Moments

Napoleon plans to attack the Allied flank from the Pratzen Heights

The Allies plan to attack across the Goldbach stream

The Allies capture Tellnitz but do not exploit the advance

Soult's attack across the Pratzen Heights collides with the Allied columns

Bagration's attack in the north is driven back

The French break through on the Heights

The rout of the Austro-Russian Army

early hours of 2 December was that in the darkness the French might slip away, and deny them their victory!

The Opening Moves

The Allied attack began at dawn in confusion. Kollowrath and Miloradavichs' forces found their advance towards Puntowitz blocked by Buxhowdens' forces, which were trying to descend from the Pratzen Heights to the Lower Goldbach. Prince Liechtenstein's cavalry, heading north towards Blasowitz, added to the delays and disruption by barging through the infantry columns. Already, Austro-Russian forces were demonstrating their inadequacies for mounting an attack that relied upon speed and coordination.

Whilst the rest of the army struggled to get moving until after 8.00 a.m., Kienmayer's detachment descended from the Heights, and at 8.30 a.m. mounted a major attack against Tellnitz. Expecting only light resistance, Kienmayer committed his forces in a piecemeal manner, with the result that in one hour the French 3rd Voltigeur Regiment repulsed five attacks. It was another hour before the lead regiments of Dokhturov's column came up and secured Tellnitz, enabling Kienmayer's cavalry to cross to the left bank of the Goldbach.

Buxhowden now intervened, ordering Kienmayer and Dokhturov to halt and wait until the other Allied columns had crossed the stream. Throughout the day, although Allied forces were to succeed in crossing the Goldbach on numerous occasions, Buxhowden's restrictive orders denied them the opportunity to exploit their gains and unhinge the French defence, which was able to concentrate its smaller forces and counter-attack individual Allied gains. The arrival of Marshal Davout's III Corps on the battlefield after 10.30 a.m. helped sustain the French defence of the Goldbach. At the same time, the vigour of Prebyshevsky's and Langeron's attacks began to wane, as they became increasingly worried by strange movements to their rear along the Pratzen Heights. The Russians were right to be concerned – the troops that they could see were French.

Earlier that morning, the sound of firing to the south informed Napoleon that the Allies were beginning a major attack against his weak right flank. The vanguard of his intended attack, formed by Soult's divisions in the valley around Puntowitz, was concealed from Allied view by fog. Because the Allies could not see the French forces, Napoleon was able to delay the attack until most of their troops had moved off the Heights. Turning to Soult he inquired, 'How long do your troops need to get to the top of the Pratzen?' When the marshal estimated twenty minutes Napoleon replied, 'Very well, then we will wait another quarter of an hour.' Just after 8.30 a.m., Soult's attack began. General Louis St Hilaire's division advanced to seize the Pratzenburg at the southern end of the Heights, whilst to his left General Dominique Vandamme led his division towards the peak of Stare Vinohrady. The French advanced at the double in columns of march, and as they passed to the right and left of the village of Pratze they emerged from the mist into brilliant sunlight reflected off the snow-covered ground. It was a moment remembered ever after as the legendary 'Sun of Austerlitz'. However, when the French reached the crest of the plateau, they did not encounter a thin screen of enemy troops, but Kollowrath and Miloradavich's column still struggling towards Puntowitz. On seeing the French troops, Kutuzov exclaimed, 'That's where we're really hurt.' Realizing the danger, Miloradavich attacked St Hilaire's leading brigade, and fierce fighting erupted across the Heights.

The Crisis of the Battle

In the north, the appearance of Austro-Russian forces surprised the French and stalled their wheel southwards. Disobeying his original orders, Prince Bagration decided to attack Lannes' V Corps between Blasowitz and the Santon Hill. The first Allied attack in this area, made by Prince Liechtenstein's cavalry, was easily repulsed by the accurate musketry and artillery fire of Lannes' troops. When Bagration renewed the attack in greater strength, Murat committed Etienne Nansouty's heavy cavalry division, which crushed Liechtenstein's first line. Alternating firepower with cavalry charges, the French steadily drove the Allied cavalry back. Then in the wake of Murat's cavalry, Lannes pushed his right wing forward to take Blasowitz, from where his artillery enfiladed the Russian infantry advancing towards the Santon. The last Russian attack was repulsed on the French left near Bosenitz. With the Allied advance checked, Lannes and Murat launched a general offensive. Just after midday, with his forces under repeated French attack, and separated from the rest of the Allied Army, Bagration ordered a retreat eastward before disaster engulfed his command. Liechtenstein's shattered cavalry filtered away southeast across the northern edge of the Pratzen Heights. With the Allied right defeated, Lannes, Murat, and Bernadotte then moved their forces southwards to relieve Soult.

Victory on the Heights

On the Pratzen Heights a vicious see-saw action was raging between Soult's troops and a larger Allied force. The heaviest fighting centred on the Pratzenburg. St Hilaire's division had managed to advance to the crest of the position, where it was soon subjected to a series of heavy Allied counter-attacks. Kollowrath and Miloradavich's main body came on against the French at bayonet point. St Hilaire's right flank was also attacked by the rear of Langeron's column, which had turned about to face the French on the Heights. But although Allied troops fought bravely – on three occasions Prince Volkonsky led his Grenadier Regiment forward against devastating enemy fire – they were unable to dislodge Soult's troops from the Heights. The ability of French units to change formation and manoeuvre in a coordinated manner enabled them to exploit the ground and smash the clumsy Allied columns with devastating firepower, eventually breaking them. As the Allied troops streamed off the Heights the French moved forward, bayoneting the wounded as they had been ordered.

As the Allied troops on the Pratzenburg broke, their final attack crashed into Vandamme's division on the Stare Vinohrady. This assault was made by the Russian Imperial Guard, which had been retreating slowly southwards in order to join the main Allied Army when the appearance of Vandamme's troops on their right flank forced Grand Duke Constantine to attack in order to secure his line of retreat. The initial Russian infantry attack smashed the first French line, but was halted by the next. With the cry 'For God, the Tsar, and Russia!', Constantine unleashed the Guard Cavalry, which cut the ill-prepared French to pieces. At this moment of crisis, Napoleon arrived on the Heights with the Imperial Guard and Oudinot's Grenadiers. Marshal Bessieres led forward the French Cavalry of the Guard, and for a quarter of an hour a murderous engagement took place. Eventually the arrival of Bernadotte's I Corps convinced Constantine to break off the action and retreat eastwards.

With the Pratzen Heights secured and the Allied right and centre destroyed, the final act of the battle of Austerlitz began. St Hilaire and the remnants of Vandamme's division, supported by the infantry of the Imperial Guard, moved south over the Heights, straight into the rear of Buxhowden's forces. The end was swift, and as fire poured into the Allied troops on three sides their commanders realized that if they did not retreat south they faced annihilation. As the Allied columns fell back in disorder, panic gripped the troops and a general rout ensued across the shallow and frozen Menitz and Satschan ponds. The Austro-Russian Army was finished as a fighting force. Napoleon had won his decisive victory.

That night Napoleon rode slowly northwards across the battlefield in silence. When he heard the cries of a wounded man he would dismount and comfort him with a glass of brandy. Three days after the battle Emperor Francis of Austria and Tsar Alexander of Russia signed an armistice with the French. Their overconfidence and lack of respect for Napoleon's military genius had led them to a humiliating defeat, and cost many soldiers their lives. A grateful Napoleon told his Grande Armée,

'My people will greet you with joy, and it will be enough for a man to say "I was at the Battle of Austerlitz", and they will reply "There stands a hero!"'

TIM BEAN

Further Reading

Duffy, C.J., *Austerlitz* (London, 1977)
Chandler, D.G., *The Campaigns of Napoleon* (London, 1967)
Esdaile, C.J., *The Wars of Napoleon* (London, 1995)
Connelly, O., *Blundering to Glory* (Delaware, 1987)

Isandlwana

22 January 1879

Dividing Your Forces

'The sad disaster which has occurred to a portion of the force under my command… has thrown back the subjugation of Zululand to an indefinite period and must necessarily entail sacrifice of men and money in far larger proportions than was originally expected.'

Lord Chelmsford

Battles, like wars, are won by taking risks, and any good commander must know when to take them. It is a very old military maxim that it is a mistake to divide your forces 'in the presence of the enemy', to deliberately make yourself weak and so to invite attack. The one circumstance to which this does not always apply is the situation in which one side cannot locate the other, but is so confident of victory that it *wants* to be attacked. This was particularly the case in 19th-century colonial warfare, in which European armies were confident of their superior firepower and military skills against often elusive enemies. This confidence was usually well placed, but sometimes it produced catastrophe, as at the battle of Isandlwana in the 1879 Anglo-Zulu War, one of the greatest single disasters in British military history. The battle cost the lives of over 1,300 men (75% of the British force), including nearly 600 seasoned Imperial regulars of 1st Battalion and 2nd Battalion, the 24th Regiment of Foot (1/24th and 2/24th Foot). Fifty-two British and colonial officers died in this catastrophic British defeat, more than at the Battle of Waterloo over six decades earlier. Although the exact causes of the Isandlwana defeat continue to be debated, all are agreed that a gross under estimation of the military capabilities of the attacking Zulu army was the primary factor in precipitating this unparalleled disaster.

Chelmsford's Original Campaign and Battle Plans

The general causes of the 1879 Anglo-Zulu War constitute well-trodden ground. The independent Zulu nation represented the one remaining major political obstacle to long-term British plans to confederate the various provinces of mineral-rich southern Africa, while Paramount Chief Cetshwayo's estimated 50,000-strong army presented a perceived potent strategic threat to the security of the whole region and overall British hegemony. The economic motivation behind the war was equally important: Zulu male labour was seen as a valuable untapped resource which would help meet the avaricious needs of the rapidly expanding southern Africa mining and farming industries. War was soon guaranteed by the imposition, in December 1878, of what is generally recognized as an impossible British ultimatum to the Zulu. This included, from the perspective of Cetshwayo and his advisers, the wholly unacceptable demand requiring the immediate total disbandment of the Zulu army and, with it, the key pillar of the highly militarized Zulu society. On 11 January 1879, on the expiry of the 30 day ultimatum, over 12,000 British troops entered Zululand to begin what was widely predicted to be an easy, rapid, and cost-effective war.

The initial campaign plans of Lieutenant General Lord Chelmsford, commander in chief of the British invasion force, were simple. The overriding main strategic aim was to bring the Zulu army to battle by striking decisively at the enemy's centre of gravity, the Zulu capital of Ulundi (or more specifically, Cetshwayo's royal *kraal* or personal home at Ondini). For this purpose Chelmsford's force was divided into five columns each mixing British infantry with other forces. Three of these would advance from widely separated points on the Zululand border with Natal (already part of the British Empire) with two columns held in reserve along the border to protect local white settlers against any Zulu breakthrough. The three attacking columns would be deployed in a slowly enveloping pincer movement towards the Zulu capital. Number 4 Column, the northernmost on Chelmsford's left, was commanded by Colonel Evelyn Wood and comprised 2,250 men, while Number 1 Column on the right was led by Colonel Charles Pearson and comprised nearly 5,000 men. The main thrust of the British invasion was Number 3 Column, led by Colonel Richard Glyn, comprising just under 4,700 men, and commanded and accompanied by Chelmsford himself.

Zulu Battle Plans

The Zulu tactical deployment was one of limited but active defence. Cetshwayo, surprised at the severity of the British ultimatum and acutely aware of overall British technical superiority, relied heavily on hit-and-run tactics, with the preferred aim of picking off isolated columns and inflicting severe enough casualties to encourage a British withdrawal. As Cetshwayo explained to his captors after the war, he hoped to be able to crush the English columns, drive them out of the country, defend the border, and then arrange a peace before reinforcements arrived from across the sea.

In some ways Cetshwayo's age-old tactical deployment of his army, based on the innovations of the great Zulu Paramount Chief Shaka 50 years before, represented a small-scale version of Chelmsford's own grand pincer strategy. Zulu battle tactics hinged upon the rapid encircling and enveloping formation known as the *impondo zankomo* ('beast's horns'). The Zulu regiments, or *impis*, generally comprising the warriors, or *amabutho,* who were younger and more able-bodied, variously occupied the two wings or horns of the crescent-shaped mass with the aim of rapidly surrounding the enemy on each flank. Meanwhile, a powerful body of more experienced *amabutho* forming the 'chest' of the beast would attack and distract the enemy head-on, with a reserve or 'loins' deployed behind as reinforcements. The crux of Zulu military success had always rested first and foremost on the element of surprise and rapid momentum, with concentrated pressure applied to both the front and rear of the enemy. Once the horns had closed and the chest had moved forward, thus trapping the enemy, the crucial focus would be upon close-quarter battle using the highly effective and deadly short stabbing spear or *assegai*, the cowhide shield, and the *knobkerrie* or wooden club. Male prisoners were rarely taken by the Zulu. Such tactics were reinforced by excellent military intelligence about the enemy. Long before the war started, Cetshwayo's spies had infiltrated Natal, and established accurate information on the strength and deployment of all three of Chelmsford's attacking columns. It was a formidable killing machine, especially when deployed against a demoralized, disorganized, or above all complacent enemy.

British Complacency

Zulu strength was greatly underestimated by Chelmsford and many of his subordinate commanders attached to the ill-fated Number 3 Column. Before this war, Britain's armies had enjoyed an unbroken run of success against her African enemies. The men of 1/24th Foot in particular consisted largely of experienced veterans who had already spent four successful years engaged in fluid short-lived skirmishes with rebels of the Xhosa people, in which Martini Henry Mark 2 rifle volley fire had been deployed at up to 1,000 yards range, with maximum destructive effect. None of these rebel African groups, however, could be in any way compared to the sheer size, organization, and fighting power of the Zulu, the most powerful army in black Africa.

Already confident of his military superiority, much of Chelmsford's pre-invasion preparations were focused on the logistical (supply) or transport problems presented by a country pitted with ravines, and lacking in any form of metalled road. For this purpose he had amassed, by January 1879, nearly 220 wagons, 82 carts, 1,507 oxen, 49 horses (excluding those of his cavalry), and 67 mules controlled by 346 conduct-ors, drivers, and *voorloopers* (the Afrikaans word for animal handlers). These wagons and livestock were deployed on a continuous and circuitous return journey between the various camp sites and main supply base at the mission station at Rorke's Drift on the Buffalo River. It was an unbalanced strategy with arguably much less emphasis placed on future tactical deployments to protect such a long vulnerable logistical chain against a mobile disciplined enemy. Moreover, Chelmsford's primitive intelligence system provided little or no information about Cetshwayo's possible military deployment or plans. Above all he was a commander under pressure: his political master, the High Commissioner Sir Henry Bartle Frere, had already exceeded the British government's instructions in promoting the war, and consequently sought a cheap and rapid victory.

The British Camp at Isandlwana

Chelmsford's dangerous underestimation of his enemy clearly emerged through his selection of the Isandlwana camp site as the major staging post for Number 3 Column's route to Ulundi. The arrangement of the centre column's camp beneath the ominous sphinx-like Isandlwana crag after its arrival on 20 January, while it was located close to wood and water supplies, immediately revealed a number of defensive vulnerabil-ities. Disregarding the advice of his experienced Boer allies, Chelmsford chose neither to entrench his camp nor laager (form a defensive circle) his 100-odd wagons. At the subsequent court of enquiry into the disaster, it was pointed out that the rocky ground was unsuitable for digging, the wagons had to be kept unharnessed and mobile for supply reasons, and both laagering and trenches were not justified by the temporary nature of the site. This vast sprawling camp was therefore left highly vulnerable to sudden attack, with the unprotected 1/24th Foot, 2/24th Foot, and additional colonial forces strung out across an almost one-mile frontage, protected only by the mountain, and lacking any physical defensive barriers to the front or flanks. It was a veritable defensive nightmare, with the two battalions' ammunition wagons positioned as much as half a mile apart. Forward intelligence was also seriously compromised by the restricted use of vedettes (mounted sentries), who were not deployed beyond the lip of the Nqutu Plateau, and were therefore unable to detect the arrival and concealment of the main Zulu army.

These defensive weaknesses were compounded by a second even more fatal decision. Responding to urgent reports of a major skirmish between a sizeable Zulu force, and his own reconnoitring force under the command of Major John Dartnell, Chelmsford made the fateful decision to divide his force without knowing the exact disposition of his enemy. At 4.30 a.m. on Wednesday 22 January, Chelmsford, accompanied by Colonel Glyn, moved several miles to the southeast of Isandlwana camp in search of what he believed to be the main Zulu army, taking with him the

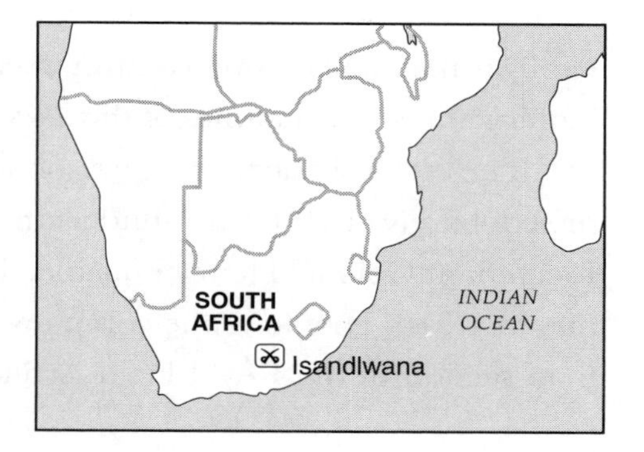

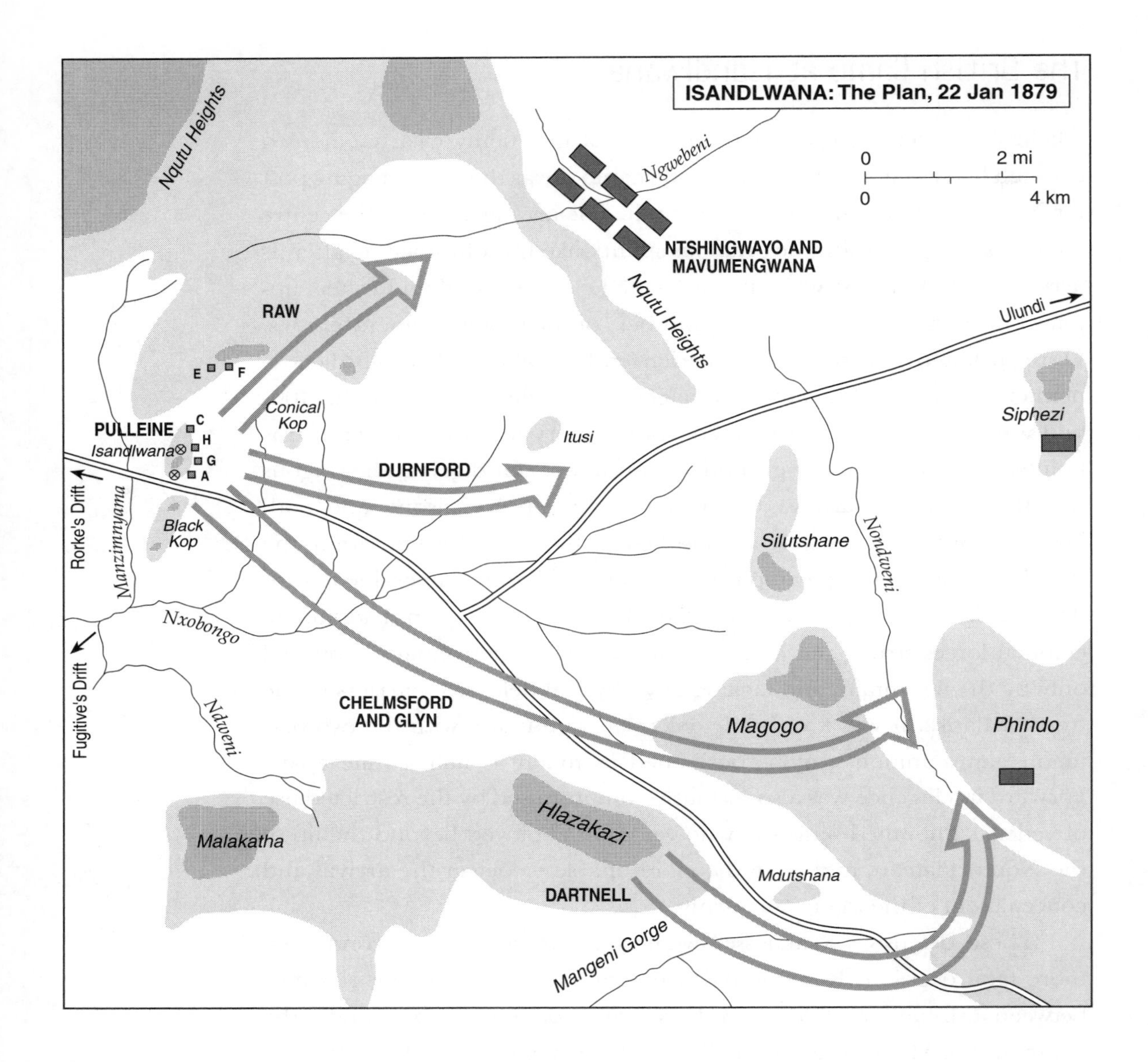

stronger half of his centre column, including four out of the six guns of the battery and six companies of the 2/24th Foot.

The seriously depleted garrison left behind at Isandlwana now comprised barely 1,700 men under the command of 41-year-old Brevet Lieutenant Colonel Henry Pulleine. The garrison comprised five companies of 1/24th Foot and one company of 2/24th Foot, supported by a 70-man section of the Royal Field Artillery with only two 7-pounder guns,

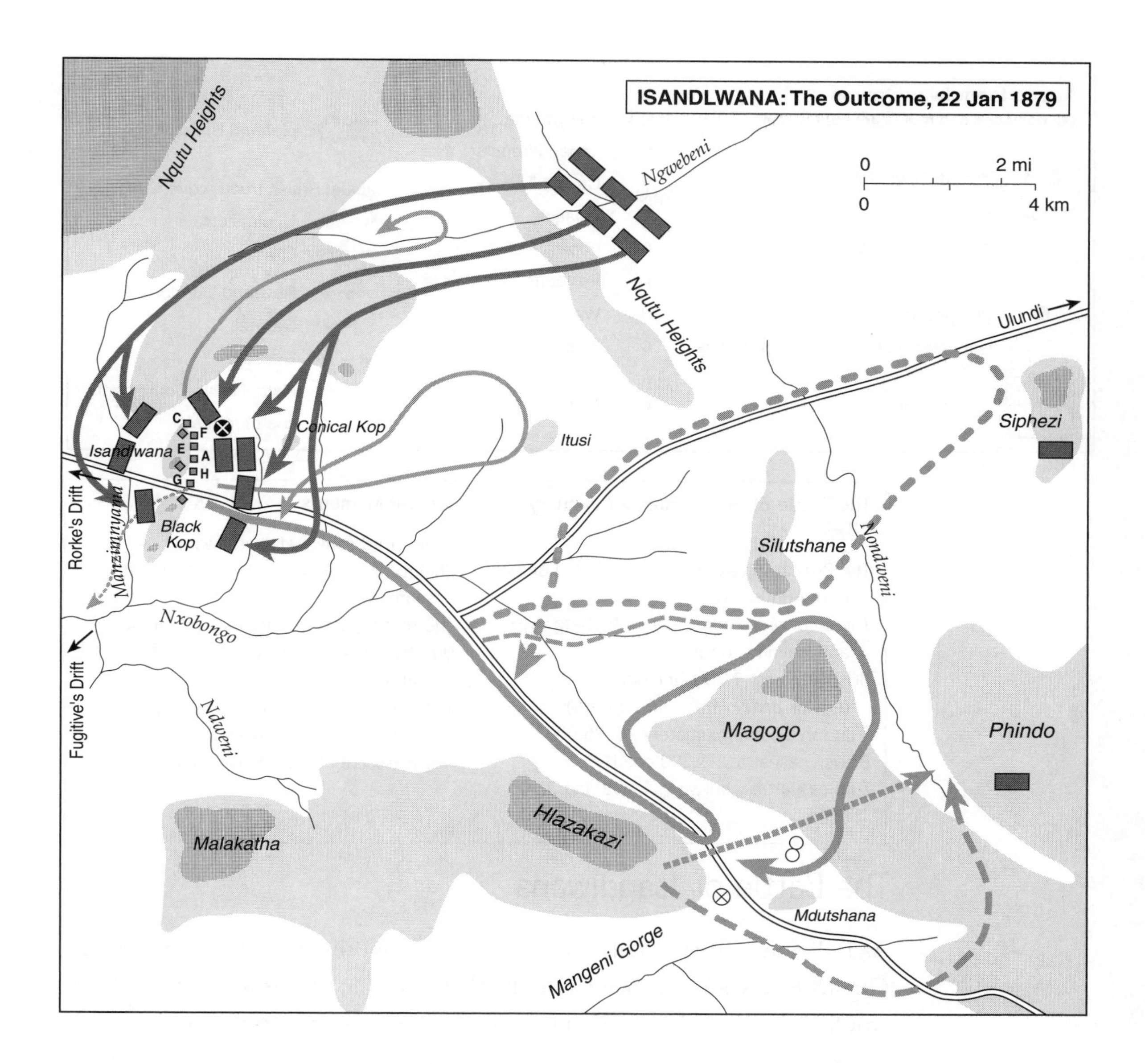

and several European and African colonial units, notably the Natal Native Contingent (NNC). Pulleine's defence of this widely dispersed camp site now ultimately rested upon strict adherence to Chelmsford's written orders: 'You will be in command of the camp during the absence of Colonel Glyn: draw in your camp or your line of defence while the force is out; also draw in the line of your infantry outposts accordingly, but keep your cavalry vedettes still far advanced.'

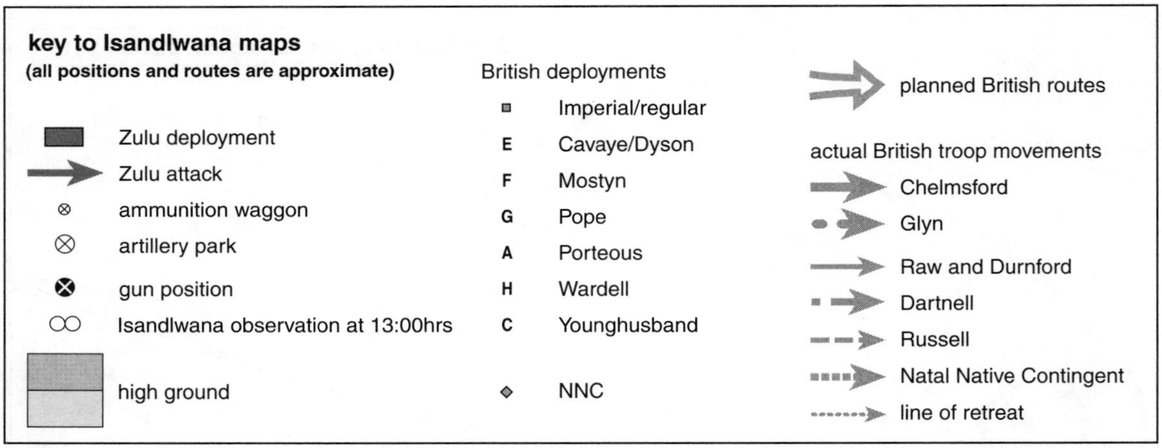

key to Isandlwana maps
(all positions and routes are approximate)

■	Zulu deployment
→	Zulu attack
⊗	ammunition waggon
⊗	artillery park
⊗	gun position
∞	Isandlwana observation at 13:00hrs
▒	high ground

British deployments

▪	Imperial/regular
E	Cavaye/Dyson
F	Mostyn
G	Pope
A	Porteous
H	Wardell
C	Younghusband
◆	NNC

planned British routes

actual British troop movements

- Chelmsford
- Glyn
- Raw and Durnford
- Dartnell
- Russell
- Natal Native Contingent
- line of retreat

The Battle of Isandlwana, 22 January 1879

The British forces under Lieutenant Colonel Henry Pulleine
The Zulu forces under Chiefs Ntshingwayo and Mavumengwana
British forces: 1,700 (including 600 regular British Imperial infantry)
Zulu forces: approximately 20–25,000
British casualties: 1,300 dead
Zulu casualties: between 1,000 and 3,000 dead

Critical Moments

Chelmsford camps at Isandlwana
Chelmsford departs to the southeast in search of the Zulu
The first reports of Zulu reach Pulleine
Durnford makes a foray to the Nqutu Plateau
The main Zulu attack on the British camp
The collapse of the British line
The massacre of the British camp

The Battle of Isandlwana

As Pulleine's vedettes failed to sufficiently scout the Nqutu heights, and as Chelmsford's force moved ever further away to the southeast, Pulleine would now face the consequences of an imminent confrontation with nearly half of the Zulu army. His fighting effectives, totalling less than 1,500 men, would be outnumbered by as many as twenty to one. The first serious sign of Zulu activity was reported in a cryptic note addressed from Pulleine to Chelmsford: 'Report just come in that the Zulus are advancing in force from the left front of camp, 8.05 a.m., H B Pulleine, Lieutenant Colonel.' Chelmsford, who was now several miles distant, again demonstrated a dangerous complacency, returning this to his junior officer, Major Cornelius Clery, without a word. When Clery asked him 'What is to be done on this report?', Chelmsford allegedly replied, 'There is nothing to be done on that.' The failure of his naval aide, Lieutenant Berkeley

Milne, to discern through his field glasses any major hostile activity in the camp area from the nearby hill only served to enhance Chelmsford's continuing overconfidence.

At approximately 10.30 a.m., the arrival at the Isandlwana camp site of Colonel Anthony W Durnford of the Royal Engineers, ordered up from Rorke's Drift and accompanied by 500 mounted African troops of the Natal Native Horse, while providing welcome reinforcements, was to compound the existing overstretch in the camp's defences. Durnford, outranking Pulleine, apparently assumed control over the camp's defences and in the absence of further orders from Chelmsford, committed his forces at around 11.30 a.m. to a foray against Zulu elements spotted on the hitherto largely unscouted Nqutu Plateau. Durnford's aggressive tactics sharply contrasted with those of the more conservative Pulleine, a long-serving officer but one noted more for his administrative achievements than his battlefield experience. Under pressure from Durnford, Pulleine reluctantly, and despite Chelmsford's earlier instructions, allowed two of the six Imperial infantry companies, led by Lieutenants Charles Cavaye, Edward Dyson, and, later, Captain William Mostyn, to deploy in support of Durnford to a position on the spur over 1,500 yards away to the left of the camp.

Just before midday, as Chelmsford's force rested several miles away in the midst of the Hlazakazi Hills, and with Pulleine's garrison troops now overextended by up to two miles of frontage, the serious consequences of Chelmsford's overall lack of intelligence about the enemy became dramatically apparent, as part of Durnford's force on the Plateau commanded by a Lieutenant Charles Raw suddenly discovered the Zulu army. The surprise was mutual; the Zulu commanders, Ntshingwayo and Mavumengwana, had not planned to attack Isandlwana until the next day but, once discovered, the lead elements of their impi now committed themselves to a full-scale attack on the weakened Isandlwana garrison.

As news of the discovery of the massive Zulu impi reached a shocked Pulleine, the remaining Imperial companies were recalled from lunch, and deployed to the left front of the camp to confront the main Zulu 'chest' (consisting largely of the uKhandempemvu and uMbonambi regiments) now cascading over the rim of the Nqutu escarpment. Three companies in camp, G Company under Lieutenant Charles Pope, A Company under Francis Porteous, and H Company under Captain

Wardell, were now hastily joined by Cavaye's, Mostyn's, and Dyson's detachments from E and F Companies, and their retreat from the spur covered by Captain Reginald Younghusband's C Company, which took position on the extreme left of the line. In these final, fixed positions the battle stabilized. All six Imperial companies had now assumed the classic battle formation for defending against a hostile attack. With the arrival of Durnford's force of approximately 200 men on the right flank, the combined volley fire from over 1,000 Martini Henry rifles and carbines supported by both the 7-pounder guns ensured that most Zulu casualties occurred at this point. For a brief while it would seem that Chelmsford's and Pulleine's battle plans were coming to fruition. Indeed, surviving eyewitnesses testified to this brief period of Imperial domination of the battlefield. Interpreter James A Brickhill remarked on the 'increasing gun roll' sustained by the regular infantry, whilst Lieutenant Horace Smith-Dorrien confirmed the initial steadiness of the Imperial firing lines. A mood of confidence, almost nonchalance, permeated through to the rank and file. Another survivor, Captain Edward Essex of the 75th Regiment of Foot, noted the 24th Foot laughing, chattering, and even joking as they unleashed volley after volley into the dense black masses.

It was a false sense of security. At one momentous point, probably some time between 1.00 p.m. and 1.15 p.m., the Imperial volley fire slackened and in some cases stopped as the companies were rapidly recalled to their tented areas. With the cessation of Imperial volley firing, retreat inevitably turned into rout, as cohesive units dissolved into desperate groups fighting for their lives. In the words of Kumbeka Gwabe, a Zulu survivor from the UmCijo Regiment, 'we spared no lives and did not ask for any mercy for ourselves.' The Imperial infantry perished almost to a man, with barely 50 whites, mostly mounted Europeans, escaping across the Saddle or Neck and along Fugitive's Drift before the two Zulu horns merged. By around 2.00 p.m. virtually all fighting had ceased, although small groups on the Drift, for example a few of the 24th under Lieutenant Edgar Anstey, possibly lasted until as late as 4.00 p.m.

Chelmsford, finally alerted some time after 2.00 p.m. by eyewitness reports, but still disbelieving the news, slowly departed from his Mangeni Gorge camp, and linked up with Commanders Rupert Lonsdale and George Hamilton-Browne with their NNC troops, who finally confirmed the terrible events. He returned to Isandlwana at dusk, to discover a devastated camp comprising a few mainly drunk or dying Zulu and over 1,300 British and allied corpses, most of which had, following Zulu fighting tradition, been stripped of their uniforms and ritually disembowelled.

The Reasons for the Disaster

The question now arises as to why the Isandlwana garrison suffered such a catastrophic collapse. Virtually all the problems encountered by Pulleine's force can be traced back to the flaws in Chelmsford's initial battle plan. First of all, the Imperial lines were clearly grossly overextended, and highly susceptible to any flanking movement. Chelmsford's departing order to defend the camp had committed Pulleine to an impossible defence of what was, in effect, a great deal of unnecessary terrain. Vastly outnumbered, the only hope for Pulleine was to sustain continuous Imperial volley fire. This in turn depended upon an uninterrupted supply of ammunition, and it was here that Chelmsford's disposition of the camp site played a crucial role in precipitating the disaster. Virtually all sources agree that the excessive distance of Durnford's mounted colonial force from the main ammunition supply was almost certainly the key factor in the collapse of the right wing. Jabez Molife, one of the few survivors from Durnford's command, vividly recalled that, after the initial check to the Zulu left horn, 'our cartridges were nearly done' with inexplicable delays in securing fresh ammunition from the camp. It is probable that the British left wing was also severely compromised by its distance from the ammunition wagons. Captain Henry Hallam Parr, who interviewed survivors of Isandlwana, confirmed that Cavaye's, Mostyn's, and Dyson's men, almost certainly fatigued from their rapid return from the spur and longest in the firing line, were 'very short of ammunition, and their initial accoutrement of only 70 rounds apiece was rapidly expended owing to the hot fire they had been forced to sustain to keep the Zulus from closing upon them while they were retreating on the camp.' At an average of three rounds per minute, one source estimates that the Imperial infantry alone expended up to 70,000 rounds during the first half hour of the main battle. Ammunition distribution was clearly not helped by the inflexible nature of quartermastering, with at least one quartermaster, Edward Bloomfield of the 2/24th Foot, recorded as obdurately refusing the issue of cartridges to non-battalion members. But it was above all the poor deployment of the ammunition wagons containing nearly half a million rounds which played a crucial role in the collapse. Lieutenant Smith-Dorrien provided decisive and damming evidence: 'the reader will ask why the fire slackened and the answer is, alas, because with thousands of rounds in the wagons 400 yards to the rear there was none in the firing line; all those had been used up.' The recorded positions and alleged panic of elements of the NNC, which may have caused the first gaps in the British line of defence, remains hotly contested today. But widespread panic on their part was quite understandable, as they were poorly equipped and generally armed with less than 30 rifles per company. In any event, the collapse of Durnford's right wing because of ammunition failure, and possibly sections of the left wing for the same reason, enabled the Zulu horns to fatally outflank Pulleine's already overstretched force.

Secondly, Chelmsford's, and to a lesser extent Pulleine's and Durnford's, failure to establish a final reserve or bastion of defence clearly played a major role in the disaster. The absence of any laagering of the wagons, and of entrenchments, meant that the troops, once in retreat, stood no chance of recovering or stabilizing their position. Had there been a final redoubt position, possibly in square formations at the base of Isandlwana Mountain or grouped around the vital ammunition stores (as later at Rorke's Drift), many more men might have survived. In the event, devoid of such physical protection and vastly outnumbered, men were left to perish alone or in small vulnerable groups.

Thirdly, Chelmsford, and again to a lesser extent Pulleine and Durnford, had clearly totally misjudged not only the mobility but the sheer raw courage, fighting spirit, and resilience of the Zulu army. Even at the height of the battle, as the uKhandempemvu, uMbonambi, and inGobamakhosi regiments in particular were decimated by British volley fire, delivered from as little as 150 yards range, there had been no Zulu mass panic and no attempt to retreat. When the Imperial lines, starved of ammunition and fatally outflanked, had finally wavered, the Zulu bravery and overall fighting spirit was graphically symbolized by one brave induna, Ndlaka of the uKhandempemvu who, before being killed by a rifle bullet to the head, exposed himself to the British firing line in order to rally his warriors for the final ferocious charge. Zulu sustained resistance and resilience to the hitherto unchallenged destructive power of the

Martini Henry rifle was undoubtedly a major contributory factor to the eventual defeat of Pulleine's force.

Finally, Chelmsford's fatal decision to divide his force early that morning undoubtedly represented a major tactical error.By 1.00 p.m., at the critical moment of the battle, Chelmsford's force was at least three hours marching time away and any possibility of rescue or reinforcement was out of the question. In the interim period, after the first message from Pulleine received at 9.30 a.m., Chelmsford had stubbornly sustained his aloof attitude to the steady procession of reports and rumours of fighting occurring within his camp area.

In mitigation, neither Chelmsford nor his subordinate commanders, in their long military experience, had ever before confronted such a formidable enemy as the Zulu. Nevertheless, the poor initial defence plans and the misjudged decision to divert to the southeast were both negligent acts bordering on recklessness. To the end of his days, even after the subsequent Court of Inquiry, Chelmsford never admitted to his direct culpability for this disastrous failure of his battle plans. Much of the blame was shifted on to the shoulders of the deceased Durnford and, to a much lesser extent, Pulleine for overextending their lines. In Chelmsford's obdurate view 'the rear was perfectly secure ... I consider that never was there a position where a small force could have made a better defensive stand.' But in other correspondence, addressed both to Sir Bartle Frere, and to his column commanders, Chelmsford tacitly admitted to his underestimation of the 'heavier numbers of Zulu' attacking his depleted force and, above all, the 'desperate bravery of the Zulu' which had been 'the subject of much astonishment'. It was a lesson subsequently well learnt by Chelmsford who, over six months later, deployed heavily laagered and armed square formations to finally destroy the main Zulu army at Ulundi.

EDMUND YORKE

Further Reading

Knight, I., *Zulu: Isandlwana and Rorke's Drift 22–23 January 1879* (London, 1992)

Laband, J. (ed.), *Lord Chelmsford's Zululand Campaign 1879* (London, 1994)

Laband, J., *Kingdom in Crisis* (London, 1992)

Morris, D.R., *The Washing of the Spears* (London, 1965)

Singapore

8–15 February 1942

The Impregnable Fortress

'We have had plenty of warning and our preparations are made and tested ... We are confident. Our defences are strong and our weapons efficient ... What of the enemy? We see before us a Japan drained for years by the exhausting claims of her wanton onslaught on China.' British Order of the Day, Singapore, 9 December 1941

One of the most characteristic effects of overconfidence in military planning is that when it cracks, it does so all the way. An enemy that previously seemed contemptible suddenly takes on the appearance of an overpowering and unstoppable conqueror, and a position so strong that it could be held forever becomes an apparent death trap for the defenders. This is particularly true of large empires, which have very often been defended by quite small forces, backed by a considerable amount of bluff. As it gradually dawns that an attacker cannot be stopped by past reputation alone, so the dilemma facing the defending commander grows ever greater.

The British View of Japan

Before December 1941, most Westerners entertained a very low opinion of the military capabilities of the Japanese. In part this was based on popular prejudice, including the racism prevalent at that time, which asserted the biological superiority of the 'white race' over all others. While it was believed that some non-European races could become fine soldiers, with the right sort of officers and the right sort of training, it was thought that this hardly applied to the Japanese. The caricatures of the time depicted them as short, bandy-legged, buck-toothed, and myopic. It was widely believed that if Japanese pilots attempted to fly above about 5,000 feet, the cells in their blood would expand to produce a form of the 'bends', a high altitude equivalent of the condition which affects deep-sea divers who ascend too rapidly.

This tendency to discount the Japanese fighting forces was also partly the result of a widespread belief that Japanese technology was markedly inferior to that of the West. British housewives all had stories of Japanese teacups in which the cup broke away from the handle when the tea was poured in, or of Japanese toys which broke at a toddler's first handling. It was believed that the label 'Made in Japan' meant that the product was cheap and nasty. If this was true of Japanese industrial exports to the West, it was thought that it must also be true of Japanese military technology. Newsreels of the late 1930s showed the Imperial Japanese Army using artillery that seemed to date from before World War I, while Japanese tanks resembled British designs outmoded in the early 1920s. Japanese metallurgy, too, was widely believed to be inferior. The aluminium used in Japanese aircraft contained very high levels of phosphorous, which meant the aircraft exploded when hit by a bullet, while Japanese steel, remarkably brittle, was prone to shatter if hit by a large enough projectile. Hence newsreels of Japan's battle fleet, which seemed so imposing, were greeted by audiences in Australian cinemas with roars of laughter and a chant sung to the tune of the Japanese national anthem: 'Tinny tinny tin tin, tin tin tin ...'.

All this was popular prejudice. It should not have affected the judgement of the British staff officers who during the 1920s and 1930s planned for the defence of Malaya and Singapore against the Japanese, although to some extent it did. But a far more significant influence on their opinions of Japanese capabilities was the Committee of Imperial Defence (CID) Historical Section's detailed analysis of the Russo-Japanese War of 1904–05. Published in three volumes between 1910 and 1920, the CID's study, the Official History of the Russo-Japanese War, was one of the most comprehensive the British military establishment had undertaken up to that time. Although Japan had become Britain's ally in 1902, the CID found much that was wrong with Japan's performance. Imperial Russia, operating at the end of the 4,000-mile single-track Trans-Siberian Railway, was at an immense disadvantage, but the Japanese had been only moderately successful. 'From a survey of the whole campaign', the Official History argued, '... it is impossible to avoid the conclusion that the Japanese armies were not at their best when driving home victory.' It was the private opinion of one of the British observers, Colonel (later General) Ian Hamilton, that the Japanese 'lacked pluck'. Amongst the reasons that Japan had agreed to peace in 1905, the Official History felt that the most important was that 'their spirit was waning, and surrenders were more frequent than they had been'. Quoting a Russian source, the authors concluded that, 'Many [Japanese prisoners] openly acknowledged that they were weary of war, and from the nature of numerous letters from Japan found on the killed and the prisoners, it was evident that this weariness was general.' The performance of the British-trained and British-built Imperial Japanese Navy had been better than that of the Army, but its activities, confined to home waters, had involved surprise attacks on and ambushes of technically inferior Russian warships. The Japanese Navy, unlike that of Russia, had failed to demonstrate an ability to operate at any distance from home bases.

The problems of defending Singapore from Japanese attack were first studied at the British Army Staff College in 1923, the year after Britain's ending (at the behest of the USA) of the Anglo-Japanese alliance. Amongst the students and directing staff engaged in the exercise were most of the military officers who were either to be involved in drawing up defence plans for Singapore over the next 17 years, or who, like the 36-year-old Major Arthur Percival, were to be tasked with the actual defence of Malaya and Singapore in 1942. The projected performance of Japan in this exercise was based largely on the findings of the Official History; as one student recalled 'it ended in a bloody massacre – for the Japanese'.

The actual military performance of Japan when its best forces, the Kwantung Army, began their expansion into China in July 1937 served to reinforce the findings of the Official History. Despite an overwhelming superiority in artillery, armour, and in the air, Japan proved incapable of defeating the armies of either Chiang Kai-shek's Chinese Nationalists (the Kuo Min-tang) or Mao Tse-tung's Communists. By the late 1930s the great majority of Japan's Army appeared to be bogged down in a hopeless attritional struggle in China. Indications of how Japan might fare against European forces came in large-scale border clashes with Soviet forces in Siberia in 1938 and 1939, in which the Japanese divisions were ignominiously defeated. Then in November 1939 apparently overwhelming Soviet forces were themselves defeated by the ill-equipped militia of Finland in the so-called 'Winter War' between the two countries. By 1940, then, Japan's military capability was ranked very low. Incapable of defeating the Chinese and defeated by the Russians, Japan was widely regarded as somewhat less effective militarily than Italy.

The British Defence Plans

The British plans for the defence of Malaya and Singapore only make sense if it is remembered that they were based on this widely shared underestimation of Japan's capabilities. For most of the 1920s and 1930s an attack on Malaya and Singapore seemed like an utterly remote contingency; indeed such defence schemes which were considered and partly implemented owed more to Britain's desire to appease anxieties in Australia and New Zealand, and to vicious inter-service conflicts, than to any apprehension felt in Whitehall. Thus it was that by 1940 three quite separate plans to defend Malaya and Singapore were in existence, a Royal Navy plan, a Royal Air Force (RAF) plan, and an Army plan.

The most visible symbol of Britain's intention to defend the region was the vast naval base, constructed on the northeastern shore of Singapore Island. Here it was secure against attack from the sea, protected by the more than 40 heavy guns, including 15-inch guns, in the forts spread along the southern coast of Singapore. Unfortunately, the base normally played host to only a handful of destroyers and light cruisers; in the plans a battle fleet was to be dispatched to the Far East only when an attack was deemed imminent.

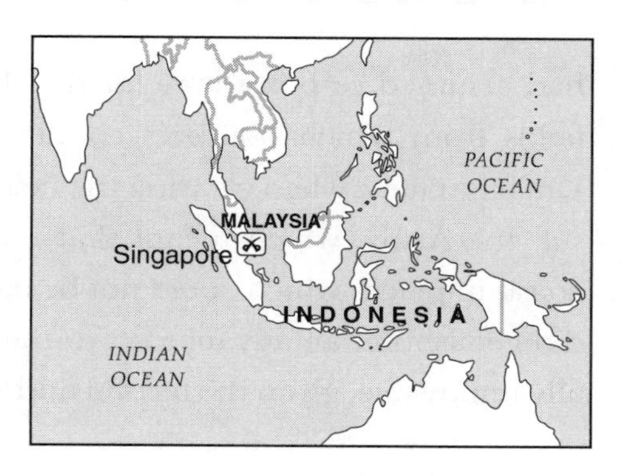

The naval base at Singapore Island was seen as invulnerable unless an enemy occupied the

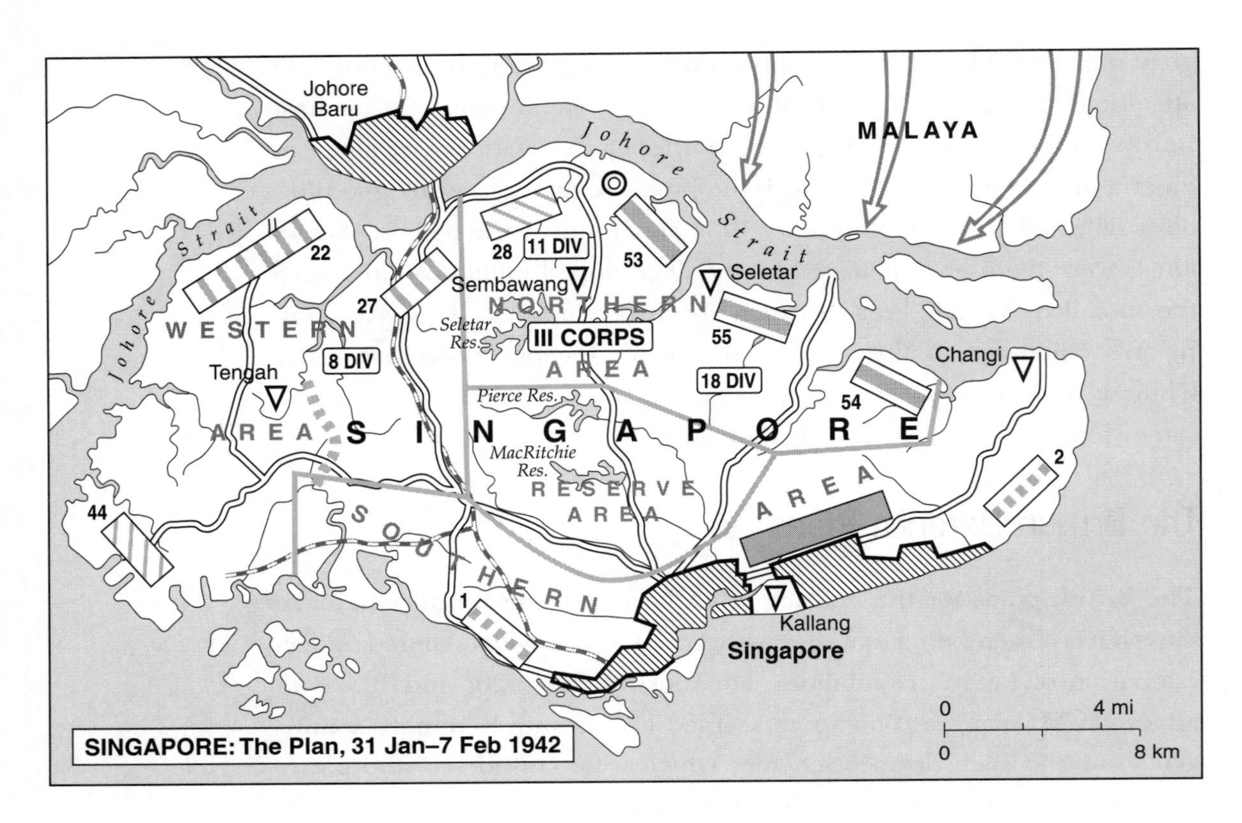

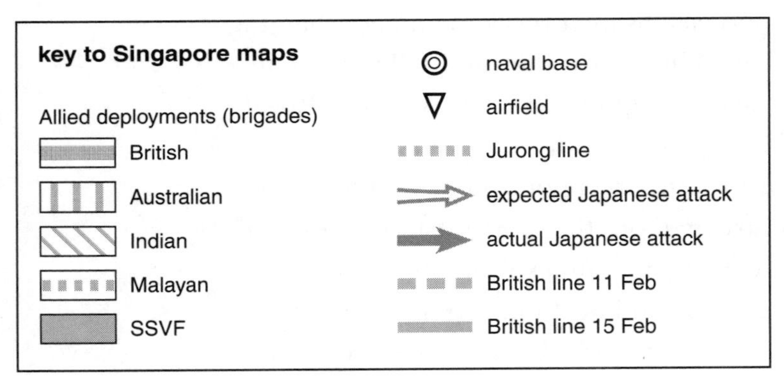

northern coast of the Johore Strait. But such an eventuality was deemed virtually impossible. Stretching more than 600 miles to the north lay the Malay Peninsula, the centre and east of which comprise jungle-clad mountains rising in places to more than 7,000 feet. The RAF had assumed responsibility for the defence of Malaya, constructing airfields from which bombers could intercept any enemy invasion fleet. Unfortunately, when clearing the fields the RAF had not bothered to consult the Army, which meant that many airfields were built close to the coast, in places which could not be easily defended. But even if an enemy did get ashore, an advance down the eastern coast was considered virtually impossible, given the terrain and the absence of communications such

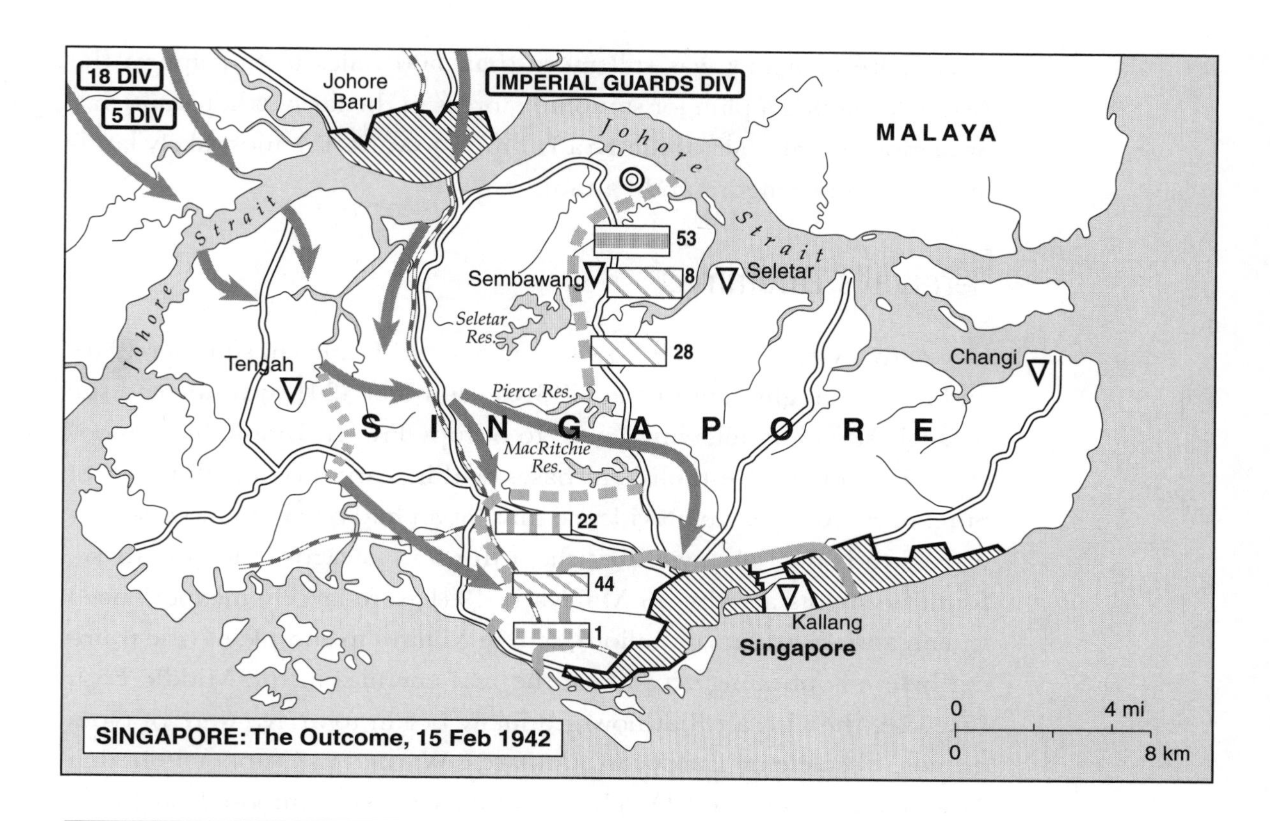

SINGAPORE: The Outcome, 15 Feb 1942

Singapore, 8–15 February 1942

The Japanese Imperial forces under General Tomoyuki Yamashita

The British Empire forces under Lieutenant General Sir Arthur Percival

Japanese forces: 30,000 troops, 168 guns, 175 light tanks, 78 bombers, 40 fighters

British forces: 100,000 troops, 400 guns, 30 armoured cars, 50 fighters

Japanese casualties: 1,714 killed, 3,378 wounded

British casualties: 95,000 (including 93,000 prisoners)

Critical Moments

The Japanese cross on to the northwest of Singapore Island

Percival fails to realize the Japanese attack is not a feint

The Australians abandon the Jurong position

Percival fails to order an effective counter attack

The 27th Australian Brigade withdraws from southeast of the Causeway

The Japanese capture the reservoirs

as roads. It was different on the western coast, a wide flat plain devoted to rice paddies and rubber plantations, where the great majority of Malaya's population lived. An Army staff report in 1937 suggested that the most feasible way of invading Malaya was to land on the eastern shores of Thailand's Kra Isthmus, and advance southwest along Malaya's western coast. Although this was still more than 500 miles to Singapore, the Army developed a plan for stationing forces in the far north, to launch a pre-emptive strike across the Kra Isthmus, towards the most likely landing beaches at Singora and Pattani.

Percival's Dilemmas

Germany's victories in the Low Countries and France in summer 1940, and her onslaught against the USSR in summer 1941, transformed the strategic situation, allowing Japan to occupy French Indo-China (modern Vietnam) and establish airbases within easy striking distance of Singapore. And yet clear evidence that Japan had begun to move southwards did not change British attitudes towards the Japanese. Reinforcements arriving in Malaya in 1941 were largely inexperienced Indian and Australian battalions, sent to Malaya to complete basic training before continuing on to face the real enemies in the Middle East. Likewise, the aircraft that now gradually began to arrive were of types deemed obsolete by European standards. Warned in October 1941 that American, British, and Dutch economic sanctions, imposed on Japan following her occupation of southern Indo-China in July, might cause Japan to strike, the British Prime Minister, Winston Churchill, at last authorized the dispatch of a Task Force of modern capital ships, the battleship *HMS Prince of Wales*, the battle cruiser *HMS Repulse,* and the aircraft carrier *HMS Indomitable*, to Singapore. So confident was Churchill that he could bluff the Japanese out of making an attack, that even when an accidental grounding prevented the *Indomitable* joining the Task Force, the Prime Minister insisted that the *Prince of Wales* and the *Repulse*, now designated Task Force Z, should sail with maximum publicity.

The British General Officer Commanding Malaya and Singapore, Lieutenant General Arthur Percival, had been promoted over 18 more senior officers when he was appointed to the Far East in spring 1941. A highly decorated infantry officer in World War I, Percival had proven a brilliant staff officer, and owed his appointment to his role in working on the defence plans of Malaya and Singapore in the late 1930s. Like Churchill, Percival did not take the Japanese threat particularly seriously. He conceived his role as doing nothing which might interfere with

Malaya's production of rubber and tin, which was now the British Empire's major source of dollar earnings. Thus he did not try to establish coherence in an impossibly muddled command structure. He had to deal with 11 different governments in Malaya, with the Governor of the Strait Settlements, Sir Shenton Thomas, and with the departments of the Malayan Civil Service, which had developed routine and procrastination into an art form. A shy, extremely polite man, Percival developed good relations with the Commander in Chief Far East, Air Chief Marshal Sir Robert Brooke-Popham, and with subordinate air and naval commanders. But he was less fortunate with his military subordinates, particularly with the commander of III Indian Corps, Lieutenant General Sir Lewis Heath, an officer of the Indian Army who was two years Percival's senior, and Major General Gordon Bennett, the outspoken and irascible commander of 8th Australian Division, who behaved as though he were Percival's co-equal and ally.

Percival knew a Japanese attack was imminent in late October, but did not seem to doubt the ability of his forces to defeat it, a confidence which was reinforced by the arrival of Task Force Z amidst a blaze of publicity on 1 December. On 7 December the Japanese made a preemptive strike against the United States' Pacific Fleet base at Pearl Harbor in the Hawaiian Islands, as the start of a general offensive intended to seize most of southeast Asia. When Japanese landings began at Singora, Pattani, and Khota Bharu on 8 December, exactly where pre-war staff studies said they would land, the first reports appearing in the press on 9 December spoke of the enemy being driven into the sea with huge losses. This was the last good news of the campaign, and it was false.

The Japanese Plan of Attack

From 10 December to 1 February, each day of the Malayan campaign brought some new calamity. It began on 10 December with Japanese land-based bombers flying from southern Indo-China locating and sinking the *Prince of Wales* and the *Repulse*, after Britain's naval commander, Vice Admiral Sir Thomas Philips, had decided to try to attack the Japanese beachheads despite his own lack of air cover. On the same day Japanese aircraft bombed and virtually destroyed the RAF base at Alor

Star, the largest in northern Malaya, their Mitsubishi Zero-sen fighters shooting the elderly RAF Brewster Buffalos out of the sky.

Two days later, a small Japanese assault force attacked the 11th Indian Division in its defensive positions at Jitra and threw it into a panicky retreat. By 18 December the Japanese advance guard had reached Penang, precipitating an undignified flight of the European population. The Japanese, under General Tomoyuki Yamashita (known as 'The Tiger'), surged on south through the new year. On 9 January a small force of Japanese tanks destroyed a potentially powerful British and Indian force at Slim River north of Kuala Lumpur, the capital of Malaya, and on the following day entered the city.

Of Yamashita's forces, the 5th Division and 18th Division were highly trained and had seen extensive service in China with the Kwantung Army, but the Imperial Guards Division had not seen action since the Russo-Japanese War. The Japanese techniques were simple. Advancing in light order, they avoided frontal attacks, feeling for and going around flanks wherever they could. They relied a great deal on psychological warfare, often sending what the British and Australians called 'jitter parties' to the enemy's rear with the expressed object of making as much noise as they possibly could – often with fireworks – in order to encourage the enemy to believe they were surrounded and thus retreat. The Japanese maintained their speed of advance by riding bicycles, which meant that unlike the road-bound British and Indians they could often move substantial forces over jungle tracks. They also lived off the land as much as they could, though huge quantities of captured supplies in Penang and Kuala Lumpur (they called these 'Churchill supplies') obviated many of their logistic difficulties.

The still confident 8th Australian Division was now thrown into action in northern Johore, with Gordon Bennett boasting that he would soon have the Japanese in retreat. But after conducting a successful ambush at Gemas on 14 January, it too was soon outflanked and falling back in considerable confusion. By 28 January, Percival had given up all hope of holding the Japanese in Johore and gave the order for all forces to retreat to Singapore, the rearguard blowing a gap in the causeway which ran from the mainland to the island. Nothing like these Japanese advances in Malaya had ever been seen before. In just 53 days, General Yamashita's forces, numbering just three divisions with about 35,000 men, had advanced nearly 600 miles at a cost of just 5,000 casualties, and inflicted 25,000 casualties (including 21,000 prisoners) on enemy forces that were numerically vastly superior.

The Defence of Singapore Island

While the Japanese advanced, reinforcement convoys had begun to arrive in Singapore, bringing some 50,000 fresh troops and more guns, while on 7 January the first of some 50 modern Hawker Hurricane fighters had flown in. By early February Singapore was awash with nearly 100,000 troops, including the entire British 18th Division. In addition, Percival had more than 400 guns, including the large fortress guns, which had been fully traversed and were now firing north, though with armour-piercing shot rather than high explosive ammunition. Supplies of ammunition were virtually unlimited, as were food supplies, with 60 giant warehouses in the dock area containing enough to feed more than a million people for many months.

Looked at in the light of his own strengths, Percival's position was far from hopeless. But having begun by underestimating the capacity of the Japanese, the succession of shocking defeats had now caused Percival to overestimate his enemy. Although British Signals Intelligence told him that Yamashita had only about 30,000 men, Percival insisted that the true number was nearer 150,000 men. Signals Intelligence also told him that Japanese convoys steaming through the South China Sea were destined for Sumatra and Java, not for Singapore, but Percival insisted on believing that a landing on the south coast of Singapore Island was a possibility. His ability to think rationally was not helped by Japanese bombing, which, though intermittent in December, became almost continuous through January, as Japanese aircraft began to operate from captured airfields. The bombing had little military significance – indeed, only one ship of a reinforcement convoy was to be sunk – but it had a very great effect on morale, in that it concentrated on Singapore's teeming suburbs, each raid causing hundreds of civilian casualties.

Percival decided that his only option was to defend the entire coastline of Singapore Island, which meant that he was unable to maintain a strong reserve in a central location. The northern shore facing the narrow Johore Strait was clearly the most likely landing point, so in the northeast, defending the now abandoned naval base, he positioned the fresh 18th Division, while the depleted Australians (reinforced by some 2,000 new recruits of questionable quality) were placed amidst the swamps and creeks of the northwest. In theory, at least, both divisions had artillery support that was more than adequate, but communication with the gunners was rudimentary in the extreme. Both divisions, too, found to their surprise that there were no prepared positions along the north coast. This deficiency had been forcefully brought to Percival's attention as early as 26 December by his chief engineer, but at that time Percival had forbidden their construction, partly because he feared it might be bad for civilian and military morale, and partly because he could not yet conceive that the Japanese would ever reach Johore. The 8th Australian Division and 18th Division now attempted to dig in, but the soldiers' efforts were frequently interrupted by artillery and machine-gun fire from across the Strait.

On 2 February, Percival gave a press conference, in which he intended to raise morale. Crowded into a long, low building, newsmen saw an exhausted, nervous man stammering his way through clearly ludicrous assertions that all had gone to plan. The kindest comment came from *The Times* correspondent Ian Morrison, who said that the performance proved 'the folly of public pronouncements unless the speaker really has something to say'. Others left the conference convinced that Percival and Singapore were finished.

Yamashita's Attack

The Japanese first assault across the Strait, launched late after dark on 8 February, came several days before the British thought it possible, since Percival's staff calculated that the movement of artillery and ammunition alone would impose delays on the Japanese. Neither Percival nor anyone else on the British side could conceive that Yamashita had attacked with only 16 battalions supported by fewer than 170 guns, each gun having only about 500 rounds apiece. Moreover, the Japanese crossed the Strait on to the northwest shore, exactly where Percival believed they would not land, and hit the three very thinly spread out battalions of the 22nd Australian Brigade. Even so, one of these, the 2/20th Battalion, inflicted huge casualties on the Japanese before it was itself outflanked, surrounded, and annihilated, showing what might have been achieved if Percival had deployed adequate forces in this area, rather than attempting to defend all points on the island's shore. By dawn the remnants of 22nd Brigade had pulled back to a switch line (a reserve defence line) which ran along a low ridge behind the village of Jurong, between Jurong creek in the south and Tengah creek in the north.

Even at this stage, the situation might have been retrieved if Percival had ordered the still unengaged 18th Division to march the eight miles from the northeast of the island to the ridge at Jurong; but he hesitated, still not convinced that the attack on the northwest was the main assault. The situation went beyond redemption when Major General Bennett prematurely sent 22nd Brigade's commander a forecasted order for withdrawal from Percival, which the brigade lost no time in implementing. By the time the mistake was realized the Japanese were on the ridge, supported by light tanks and artillery, though they were now subjected to terrifyingly heavy British bombardments.

After nightfall on 9 February, the Imperial Guards crossed the Strait at Kranji on the north coast, straight into the machine guns of the 27th Australian Brigade. The Japanese who made it ashore had to hide in mangrove swamps, pinned down by intense fire, while others were incinerated by sheets of blazing oil drifting from the bombed fuel tanks of the now abandoned naval base. At the very point of victory, 27th Brigade began to withdraw, its commander having received a garbled message from Bennett which he interpreted as an order to abandon his position and pull back closer to Singapore. Anglo-Australian incompetence had intervened decisively to save the Imperial Guards.

The Japanese now moved inland from both the north and the west, threatening the big depots and reservoirs in the centre of the island. General Sir Archibald Wavell, who had replaced Brooke-Popham as commander in chief Far East in January, flew into Singapore for the last time on 10 February for conferences with Percival and Bennett. Japanese bombers arrived during the discussion at Bennett's headquarters near Holland Road to the northwest of Singapore, destroying Percival's staff car and forcing the generals to take cover underneath a table. Wavell later recorded that he no longer felt 'much confidence in its [Singapore's] continued resistance'; that evening during a farewell visit to Sir Shenton Thomas he sat in the Governor's sitting room, thumping his knees with his fists and repeating 'It shouldn't have happened.'

The British Surrender

Had the British but known it, Yamashita's situation was now desperate. The crossing of the Strait and subsequent exploitation had cost the Japanese some 5,000 casualties out of a total force of just 30,000, and some of the Japanese batteries had run out of ammunition. Believing himself no longer capable of winning through force of arms, Yamashita determined to bluff the British into capitulation. Ordering his gunners to keep firing as though their ammunition supplies were unlimited, on the evening of 11 February Yamashita sent an ultimatum to Percival.

By this time the British commander was concentrating on his own difficulties, not those of the Japanese. The irony of the Malaya–Singapore campaign is that whereas British pre-war planning had been based on absurd underestimations of Japanese capabilities, the exact opposite was now the case. The belief that Yamashita had at least 150,000 men, unlimited ammunition, large amphibious forces in the offing, and massive air superiority (only the last of which was true) served to convince Percival of the hopelessness of his own situation. With an ever-shrinking perimeter, his forces deserting in everlarger numbers and some rioting drunkenly in the streets of Singapore, and with his own subordinates advising surrender, Percival seized on the Japanese capture of the city's reservoirs as the excuse he needed. The fact that the Japanese had not, and indeed could not, cut off the water supply was immaterial. On the afternoon of Sunday 15 February Percival surrendered Singapore to Yamashita.

In less than two months the Japanese had been transformed in British eyes from comical monkey-men to supermen possessing near demonic powers. It would take another two years' hard fighting before the British and their Allies could once more tackle the Japanese with the confident expectation of winning; while the loss of Singapore was a blow from which the prestige of the British Empire was never able to recover.

DUNCAN ANDERSON

Further Reading

Elphick, P., *Singapore: The Impregnable Fortress* (London, 1995)

Holmes, R. and Kemp, A., *The Bitter End: The Fall of Singapore 1941–42* (London, 1982)

Kinvig, C., *Scapegoat General: Percival of Singapore* (London, 1996)

Lodge, A., *The Fall of General Gordon Bennett* (London, 1986)

Narrow Margins

'No one can guarantee success in war, but only deserve it.' Winston Churchill

When a battle is lost, it is little comfort for the defeated side that their plan was a good one which could have worked, and that their army fought hard and well. But sometimes you can do your best and it just is not enough. History, at least, tends to be kind to commanders whose plans were not in themselves bad ones, but who were defeated because of things which genuinely could not have been foreseen, or because a tiny advantage swung the battle the other way. But for the winning side, even if it wins by the narrowest of margins, it is still a victory.

Gettysburg

1–3 July 1863

The High Tide of the Confederacy

'All this has been MY fault – it is I that have lost this fight.' Robert E Lee

On 6 May 1863, the evening of his victory at Chancellorsville, Robert E Lee stood at the pinnacle of his glittering military career. The first two years of the American Civil War (1861–65) in the Virginia theatre had seen almost nothing but Confederate victories, and Lee now looked for a final decisive battle to end the war successfully. Yet Chancellorsville was to lead Lee to a course of action that a little over two months later was to result in the worst defeat of his career, a battle that was a mighty nail in the coffin of the Confederacy. Like Napoleon at Waterloo, Lee's defeat at Gettysburg was with hindsight both fatal and also avoidable. The political consequences of a Confederate victory have been often debated, but it is fair to say that the history of the United States, and of the world, would have been very different if Lee had won. There were moments in the battle of Gettysburg when all this hung on the actions of only a few hundred men.

The Possibilities for the Confederates

As the Union Major General Joseph ('Fighting Joe') Hooker's Army of the Potomac withdrew to lick its wounds, the ever-aggressive Lee believed that the time was ripe for a fresh invasion of the North. His previous incursion into the territory of the United States had resulted in the bloody clash at Antietam (or Sharpsburg) on 17 September 1862, a tactical stalemate but a strategic defeat for Lee, who was forced to pull back to Confederate soil.

This time, Lee was convinced that things would be different. But Lee was just one of a number of commanders competing for both scarce resources and the ear of Jefferson Davis, the Mississipian President of the Confederacy at his capital in Richmond, Virginia. Both the Western and Tennessee theatres were vital to the Confederacy, and two senior officers, Lieutenant General Pierre G T Beauregard and Lieutenant General James ('Old Pete') Longstreet, separately proposed making a major effort in Tennessee.

Lee would have none of it. One of Lee's major defects as a commander was his excessive parochialism. A Virginian, who had only joined the Rebel cause out of loyalty to his native state, Lee underestimated the gravity of the strategic situation in the West, firmly believing that the war would be won or lost in the Eastern theatre. Lee would have to deplete his army to reinforce any attack in the Western theatres. He was not prepared to countenance this, nor was he prepared to lead an offensive in Tennessee, as Beauregard suggested. Eventually, Lee won the argument. Arguably, in doing so, he lost the war for the South.

On 3 June 1863 Lee set his Army of Northern Virginia in motion from Fredericksburg, heading for the Shenandoah Valley. Lee's army of 75,000 men now consisted of three corps, under Longstreet, Lieutenant General Richard S Ewell, and Lieutenant General A P Hill, and a cavalry division under Major General J E B ('Jeb') Stuart. On 9 June Stuart's horsemen clashed indecisively with Union cavalry at Brandy Station, in the largest cavalry action of the Civil War. As ever, the Confederate advance from Fredericksburg was skilfully handled, and Hooker marched in parallel to the Rebels, ensuring that the Army of the Potomac remained interposed between Lee and Washington. Ewell's Corps forced a Union detachment out of Winchester on 14–15 June, and then Lee's army advanced across the Mason-Dixon Line into Pennsylvania. On 13 June Hill's Corps, which had stayed in Fredericksburg keeping watch on the Union forces, also advanced north after Lee.

Lee's Strategy

The invasion of Pennsylvania apparently offered a number of prizes to the Confederates. It would maintain the momentum of the success at Chancellorsville, and would forestall the otherwise inevitable fresh Union advance into Virginia. It offered a chance to allay the logistic (supply) problems of the army by living off the land of the enemy. However, Lee was all too aware that these could only be fleeting advantages, for his army was incapable of maintaining itself on Northern soil indefinitely. Lee's invasion was in reality merely an incursion or raid. The one thing that would bring lasting success was a decisive Confederate victory that would lead the war-weary and success-starved Northern public to demand peace, and perhaps even prompt the British and French to intervene in the war on the Confederate side. Lee planned to combine the strategic offensive with the tactical defensive. He believed that the Army of the Potomac would be forced to attack the invading Rebel army somewhere north of Washington. This would allow Lee to stand and fight on ground of his own choosing – providing, of course, that an out-generalled Hooker had not already once more offered up his army like a lamb to the slaughter.

Lee could have pursued an alternative defensive strategy, remaining in Virginia, hoping to wear out the Northern armies; but he chose instead to gamble on crossing the Potomac. In the words of the British military critic J F C Fuller, Lee 'rushed forth to find a battlefield, to challenge a contest between himself and the North'. On 25 June Lee made a fateful decision that contributed mightily to his failure to win a decisive victory in Pennsylvania. Jeb Stuart was still smarting after the affair at Brandy Station, and gained Lee's permission to leave the main body of the army and try to get around the rear of Hooker's army. Such a bold use of cavalry in an independent role had worked before, notably when Brigadier General Earl Van Dorn had attacked and destroyed the Union depot at Holly Springs, Mississippi in December 1862, thus wrecking Major General Ulysses S Grant's first advance on Vicksburg. But this time the gamble was to go badly wrong. In dispatching Stuart's cavalry, and giving him such a wide degree of discretion, Lee deprived himself of his main means of reconnaissance. If not entirely blind, the Army of Northern Virginia advanced into Pennsylvania suffering from acute myopia.

Hooker's and Meade's Strategy

Hooker's immediate reaction on hearing of Lee's advance up the Shenandoah Valley was to propose to move *south*, and cross the Rappahannock. Overruled by President Abraham Lincoln, Hooker then proposed an advance directly on Richmond. Once again Lincoln overruled him, the civilian former lawyer from Illinois revealing a greater grasp of strategy than the hard-bitten soldier: 'I think *Lee's* army, and not *Richmond*, is your true objective point', Lincoln wrote, and on 25 June the Army of the Potomac started to move north after Lee. Hooker had compounded his lacklustre performance at Chancellorsville by passing up the opportunity to attack Lee while his army was in line of march. Thus on 27 June Lincoln and Major General Henry W Halleck, the Army Chief of Staff, accepted Hooker's resignation, which he had tendered over a trivial matter, and appointed Major General George G Meade in his place. Meade did not want the position, but accepted the challenge loyally. He continued the move of his army north of the Potomac, and advanced to within striking distance of Lee's flank.

Lee was now paying the penalty for keeping Stuart on a very loose rein. Stuart's cavalry had found Union resistance tougher than they had expected, and had been drawn further and further away from Lee's main body. Thus Lee, deprived of his outer reconnaissance screen, did not discover until 28 June that Meade had concentrated his army at Frederick, Maryland. Lee made haste to concentrate his scattered corps and advance east and south to meet the Union army. Cautiously, Meade moved north, towards the small town of Gettysburg. On the last day of June, 1863, Union Brigadier General John Buford's cavalry division took up positions in Gettysburg. Some time later, Brigadier General James J Pettigrew led a force of Confederate troops to seize a supply of shoes believed to be in the town. The two forces clashed, and the overture to the battle of Gettysburg began.

The First Day at Gettysburg

Gettysburg is a classic example of what is called a 'meeting engagement' or an 'encounter battle'. Neither side had planned to fight a major battle there; it just happened. In Lee's words, 'A battle thus became, in a measure, unavoidable.' Lee's enthusiasm for an offensive battle was not shared by Longstreet, who favoured the adoption of a defensive stance, getting between Meade and Washington and forcing the Union army to attack. Indeed, Longstreet was to claim that in attacking at Gettysburg, Lee went back on an earlier promise not to 'assume a tactical offensive, but to force his antagonist to attack him'. Lee's decision to attack the enemy at Gettysburg is one of the most controversial aspects of his career.

The terrain at Gettysburg strongly favoured the defender. The town lay on low ground between two minor streams, Rock Creek and Willoughby's Run, the eastern and western branches of the River Monocacy. A ridge of high ground ran between the town and Willoughby's Run, the section immediately west of Gettysburg known (from the Lutheran establishment situated on it) as Seminary Ridge. To the west lay McPherson's Ridge, and further beyond that Herr Ridge. A second, shorter spur of high ground known as Cemetery Ridge ran to the south of Gettysburg, with the commanding height of Cemetery Hill at its northern end. Less than three miles south of Gettysburg the ridge ended, with a peach orchard three quarters of a mile west of its southern end, and a very rocky area known as 'Devil's Den'. Two detached hills that have gone down in history as Little Round Top and Round Top completed Cemetery Ridge to the south. Half a mile to the east of Cemetery Hill lay Culp's Hill, by Rock Creek, with Power's Hill a mile further south. Gettysburg was a key communications junction, and ten important roads passed through the town, including the Cashtown road running to the northwest and the Fairfield road to the southwest.

Buford's cavalry repelled Pettigrew's men west of Gettysburg on 30 June, but on 1 July at 5.30 a.m. two divisions of Hill's Corps (Major General Henry Heth's and Major General William Pender's) clashed with Union troops as they advanced down the road to Gettysburg from Chambersburg. Lee's army was still suffering from two major defects: it was not properly concentrated, and the dispositions of the Union army were still far from clear to Confederate high command. Situated on the high ground to the west of Gettysburg, with outposts along the Cashtown and Fairfield roads, Buford's mostly dis-mounted cavalry div-ision had the classic role of the advanced guard: hold until relieved.

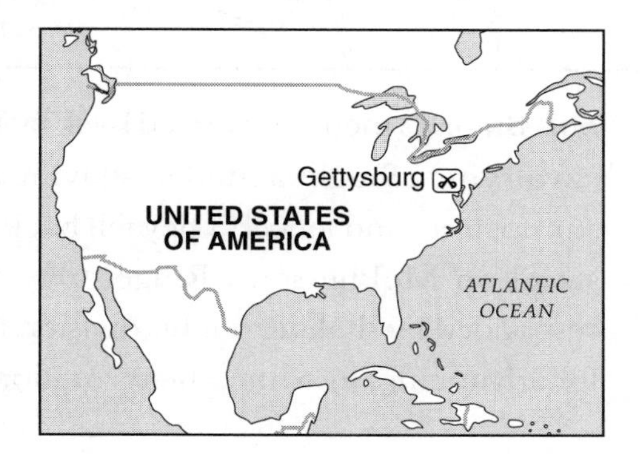

The first casualty of 1 July 1863 may well have been the pet dog of the 5th Alabama Regiment, wounded by Union fire. But the battle rapidly became more seri-

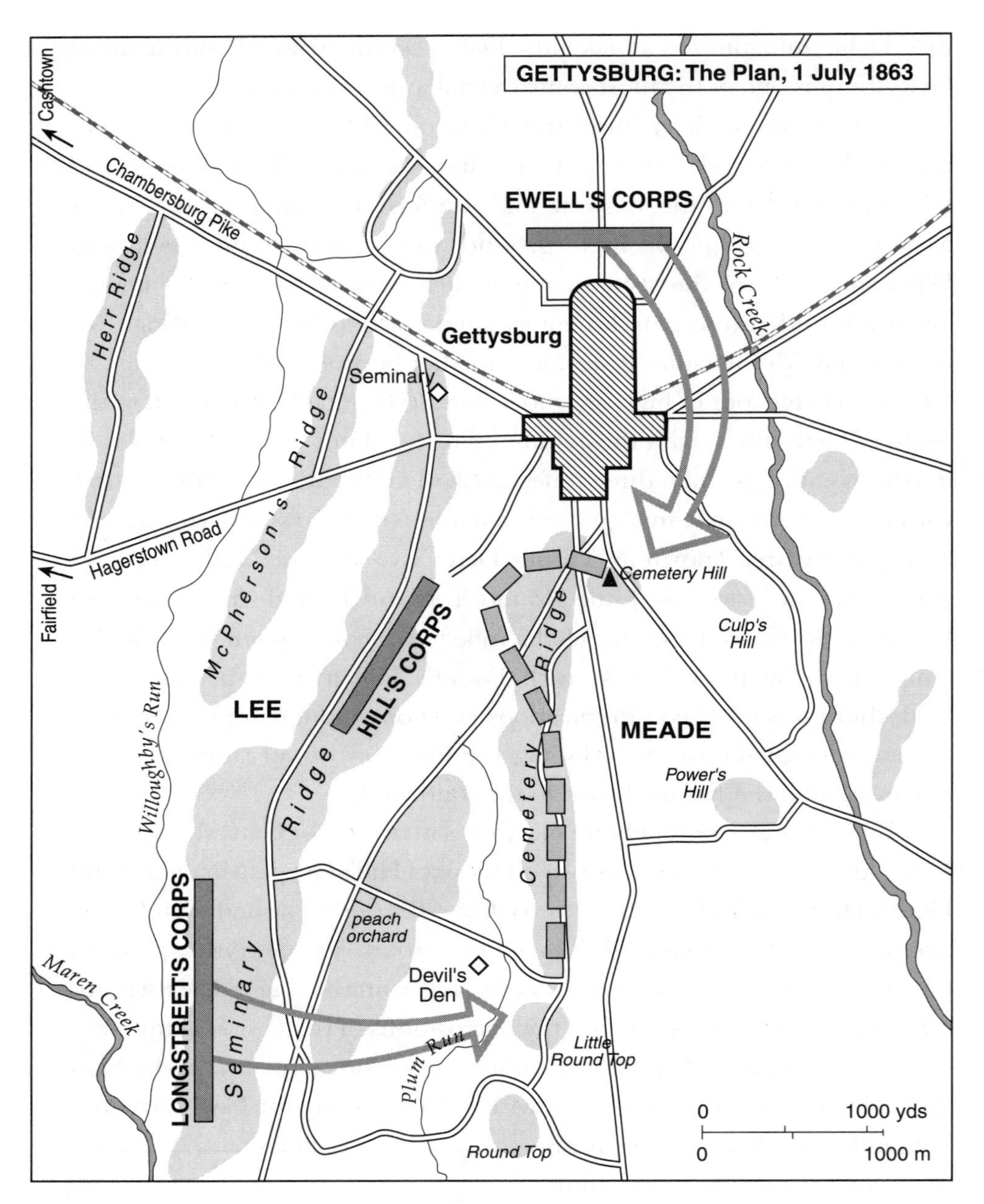

ous. Buford's men were forced back by Heth's advance, but the dismounted cavalrymen fought a model delaying action, making use of 'every stone, stump, tree, and fence'. They fell back over Herr's Ridge but then made a stand on McPherson's Ridge. On seeing Buford's men deploy, Heth responded by shaking out his division from column, the correct formation for advancing, into line, the formation for fighting. At 10.30 a.m. Buford

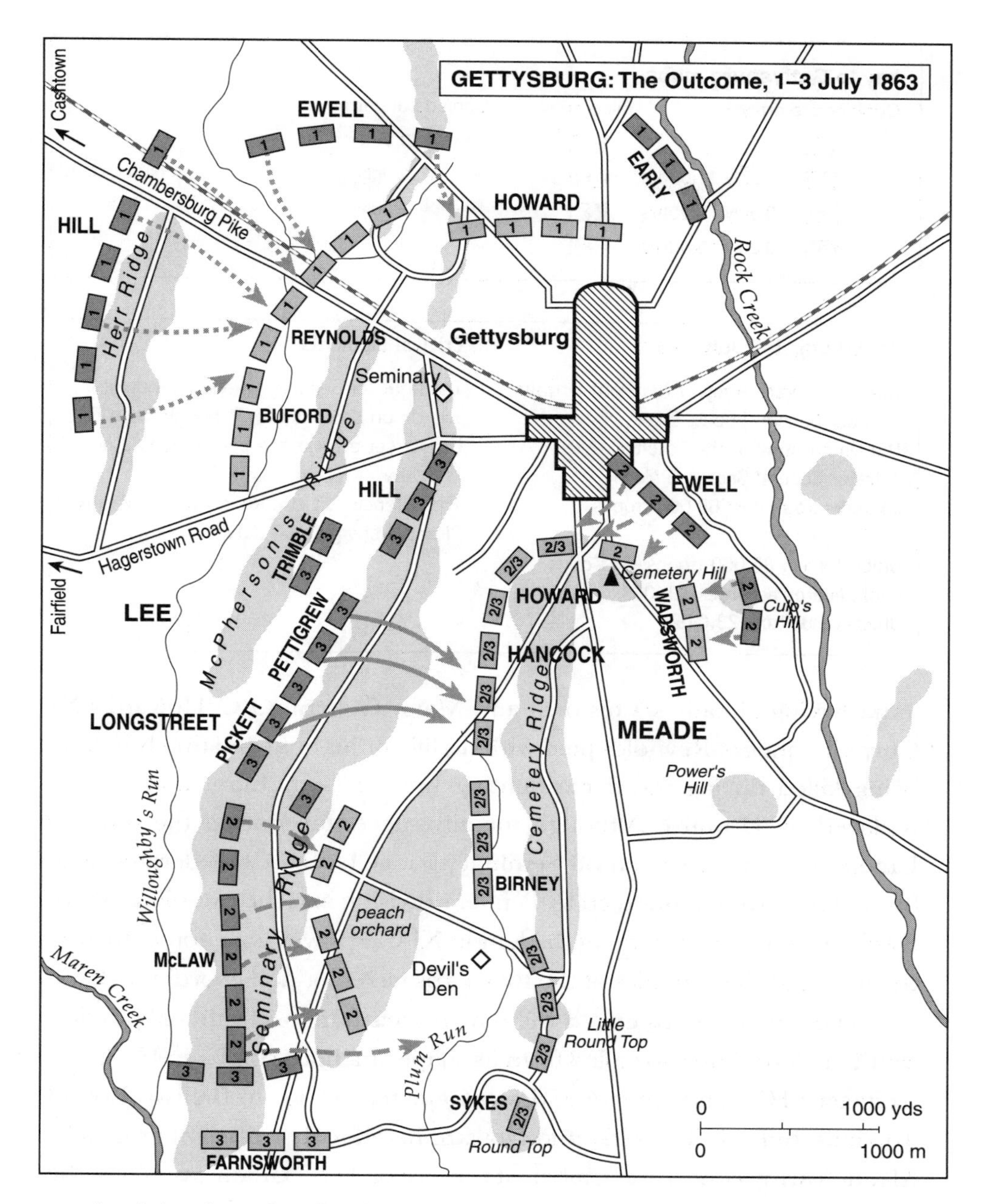

received timely and welcome help, as Wadworth's Division of the Union I Corps arrived, with the rest of the corps behind it (Union corps had numbers, Confederate formations were always known by their commanders). Heth's Confederates were still two miles short of Gettysburg.

Major General John Reynolds, commanding I Corps, was well aware of the stakes at issue. He had reached Buford's position at about 10.00

key to Gettysburg maps

Confederate army		Union army	Confederate army attacks	
▭		▭	1 July	high ground
1	1 July 11:00hrs	1	2 July	
2	2 July 15:00hrs	2	3 July	
3	3 July 15:30hrs	3		

Gettysburg, 1–3 July 1863

The Confederate Army of Northern Virginia under General Robert E Lee
The Union Army of the Potomac under Major General George G Meade
Confederate forces: 65,000 troops, 272 guns
Union forces: 85,000 troops, 354 guns
Confederate casualties: 28,063
Union casualties: 23,049

Critical Moments

The fight for Seminary Ridge on 1 July
The Union defends the 'Fishhook' position
Longstreet attacks the peach orchard on 2 July
The defence of Little Round Top on 2 July
Pickett's charge on 3 July

a.m., having already set his own and Major General O O Howard's XI Corps in motion. Reynolds paid with his life for his bold and timely action, being killed during the fierce fighting west of Gettysburg, and he was replaced by Howard. Although initially successful, during the day the Union troops were driven off Seminary Ridge by the Confederates, who by 1.30 p.m. were reinforced by Pender's division and the two div-isions of Ewell's Corps. The arrival of the Union XI Corps was not enough to hold the line, and the activities of the two corps were not well coordinated.

The Union troops ended the day in their famous 'fishhook' position south of Gettysburg, so called from its shape, anchored on Culp's Hill and Cemetery Hill, and extending down Cemetery Ridge to the two Round Top hills. Buford, Reynolds, and Howard had bought sufficient time to let Meade's army concentrate, albeit at the cost of 9,000 Union casualties (the Rebels suffered 6,500 on the same day). On the Confederate side, Ewell was armed with discretionary orders from Lee to take Cemetery Hill, 'if he [Ewell] found it practicable'. There was still plenty of daylight, and some have argued that if Ewell had pressed ahead boldly with an attack against the Union positions at about 5.30 p.m., 1 July 1863 would have ended with a substantial Confederate victory. However, ultimately Lee himself, who had arrived on the battlefield at midday and who was on

Seminary Ridge in the early evening, must carry the responsibility for the failure to take Cemetery Hill. Lee decided – probably wisely, given the tactical situation – against giving Ewell a direct order to attack. Thus the scene was set for the battle to be renewed on the next day.

Lee's Plan of Attack

During the night, Union and Rebel formations poured onto the battlefield. By morning on 2 July, Meade had 51 brigades of infantry supported by 7 cavalry brigades and 354 guns. Lee had only 34 infantry brigades and one cavalry brigade, plus 272 guns. The Union army occupied three miles of the fishhook position, more than 15 defenders to every yard of front held. Lee's troops, spread out along the circumference of a crescent-moon-shaped position, could muster only six men to the yard.

For the first time in the battle, Lee could make a proper plan. Longstreet was in favour of an attempt to move round Meade's southern flank and interpose the Army of Northern Virginia between the Federal forces and Washington. This would, in keeping with Lee's original concept of the campaign, hand the Confederates the advantage of both the strategic offensive and the tactical defensive. Lee rejected this plan, perhaps worried about his lack of cavalry to screen such a move, and instead decided to send Longstreet's Corps to attack Meade's southern flank. At the other end of the battleline, Ewell was to support Longstreet's main assault with a diversionary attack on Cemetery Hill from the north.

Longstreet's Corps did not attack until 4.00 p.m. A combination of delays in moving troops up, Longstreet's lack of confidence in the plan, and what Clausewitz often called the 'friction of war' were largely to blame. But whatever the cause, Meade used the breathing space to reinforce his line. Three Confederate divisions, from left to right, Anderson's, McLaws's, and Hood's, fell on the Union defenders. A fierce struggle developed around the peach orchard and Devil's Den, where Major General Daniel Sickles had on his own initiative moved his III Corps ahead of the main position. While this proved disastrous for the men of his corps, it also meant that the Confederates were faced with making a frontal assault. Union reinforcements joined Sickles's men in a fruitless attempt to hold back the Rebel tide. As a result, the Union troops temporarily abandoned Little Round Top, and having helped to push back Sickles, Hood's Southerners advanced on the hill. If they could take it, the entire Union line would be imperilled. The battle was in the balance.

The Crisis at Little Round Top

Two men saved Little Round Top, and probably the battle, for the Army of the Potomac. The first was Brigadier General Gouverneur K Warren, Meade's chief of engineers. Finding the hill occupied only by a handful of signallers, Warren urgently summoned up reinforcements and personally assisted two guns to be moved up to the summit. The Union forces reached the rock-strewn wooded heights just ahead of the Rebels, and repulsed the Confederates. The second saviour of Little Round Top was Joshua Lawrence Chamberlain, a college professor who was now the colonel of the 20th Maine Regiment. This 386 men strong volunteer unit, manning the vital left flank of the Union line, withstood two hours of attacks from the Texans and Alabamians of Law's Brigade, suffering heavy casualties in the process. At one stage, on the point of being outflanked, the regiment carried out a parade-ground manoeuvre and reformed into an L-shaped line; a remarkable feat, given that it was done on broken ground and under heavy fire. Still more remarkable was what happened later when the Alabamians were massing for yet another attack. The 20th Maine had to hold Little Round Top at all costs, yet had lost about one third of its strength and was running out of ammunition. Chamberlain took the only option left: to fix bayonets and charge!

Like a cobra uncurling, as the bent-back left wing swung to face the front, the 20th Maine started down the hill. Taken utterly by surprise, the Alabama troops were pushed down the slope. Suddenly, Company B of the 20th Maine, who were acting semi-independently as sharpshooters, emerged as if by magic and opened up on the rear of the Confederates. Exhausted, surprised, and now seemingly surrounded, the Alabamians broke and tumbled down Little Round Top. The pursuing Maine infantry took 400 prisoners. In this action they had lost 130 men killed or wounded and inflicted 150 casualties on the men from the Deep South. Little Round Top was saved. Lee's fleeting chance to win the battle on 2 July vanished. Ewell's evening attack on Cemetery Hill also failed. On 2 July both sides had incurred another 9,000 casualties.

Lee's Last Attempt at Victory

The events of 3 July, although dramatic, form little more than a coda to this story of Gettysburg. Lee abandoned any thought of manoeuvre. Instead, over Longstreet's heated protests, he ordered some 15,000 infantry of Longstreet's and Hill's Corps to make a frontal assault on the centre of Meade's position. Unlike the events of the first two days of battle, when Confederate plans came close to succeeding, this attack – known as 'Pickett's Charge' after Brigadier General George Pickett, one of Longstreet's divisional commanders – was doomed to failure from the start. A deafening artillery bombardment failed to dislodge the defenders on Cemetery Ridge, who greeted the rebel charge with quite literally devastating artillery and rifle fire. Perhaps 5,000 Confederates got to their objective, before they were forced back, killed, or captured. With Lee's army defeated, the battle was over and the Union had won.

Gettysburg is a classic example of sound plans being thwarted by brave and timely action on the other side. Lee's advance into Pennsylvania might well have brought him the battle north of Washington that he wanted: a defensive action, where he could smash the Army of the Potomac at his leisure. Presented with an unforeseen encounter battle at Gettysburg, Lee was denied victory by the holding action of Buford, Reynolds, and Howard. On 2 July, his attack on the Union left came close to succeeding: but Warren and Chamberlain deserve the major share of the credit for holding Little Round Top. By contrast, Lee's plan for 3 July was deeply flawed. His acceptance of the blame for the Confederate defeat at Gettysburg was both morally cour-ageous and accurate.

G D Sheffield

Further Reading

Foote, S., *The Civil War : A Narrative, Volume III* (London, 1991)
Gallagher, G.W., *Lee the Soldier* (Lincoln NA, 1996)
Hattaway, H. and Jones, A., *How the North Won* (Chicago, 1991)
Pullen, J.J., *The Twentieth Maine* (Dayton, 1991)

The First Battle of Ypres

31 October 1914

The Narrowest Margin

'The King seemed very cheery but inclined to think that all our troops are by nature brave... I told him of the crowds of fugitives who came back down the Menin Road from time to time during the Ypres Battle having thrown away everything they could, including rifle and pack, in order to escape, with a look of absolute terror on their faces, such as I have never seen before on a human being's face.' General Sir Douglas Haig

It is no easy judgement to call a battle won, if victory marks the start of a long and costly war. But what if that long and costly war leads eventually to triumph? On the outbreak of World War I, all sides hoped for a short and successful campaign, 'over by Christmas'. In summer and autumn 1914 the German Empire, with its attack on neutral Belgium and on France, came closest to success, but its failure to break the Allied lines led instead to the stalemate of the Western Front, and ultimately to Germany losing the war. In the central English cathedral city of Worcester, a pretty expanse of grass and flowers called Gheluvelt Park commemorates the action of men of the 2nd Battalion of the Worcestershire Regiment ('2nd Worcesters'), whose bravery at a moment of crisis meant the difference between that defeat and quite possibly a German victory.

The Rival Strategic Plans

The British officially date the First Battle of Ypres as lasting from 19 October to 22 November 1914, and even this was part of a much larger series of battles fought to prevent a breakthrough by either side on the northern flank of the Western Front. In September the British Expeditionary Force (BEF) under Field Marshal Sir John French had played an important part in helping the much larger French forces stop the Germans at the Battle of the Marne. Both sides then attempted to turn the open flank in what became inaccurately known as the 'race to the sea'. In October the BEF began to transfer by rail to the area of Ypres (known to the British with their usual disregard of other languages as 'Wipers'), partly to be closer to their communications through the English Channel ports, and partly to absorb additional British forces which were falling back from Ghent and Antwerp, where they had fought alongside the Belgians.

There was no overall Allied commander for the First Battle of Ypres, but there was at least a common battle plan, agreed on 10 October between Field Marshal French and the commander of the French forces in the area, General Ferdinand Foch. The Allies intended to advance in unison from the Ypres area in a northeasterly direction deep into Belgium, in order to trap the German III Reserve Corps – which had been besieging Antwerp and was now advancing along the Belgian coast – by separating it from the Sixth Army under Crown Prince Rupprecht of Bavaria, and so turn the entire German line from the north.

This plan was based on the assumption that III Reserve Corps constituted the only sizeable German force north of Ypres. Unfortunately this was completely wrong. Just as the BEF was being moved northward by train, so at the same time the four corps of German Fourth Army under Duke Albrecht of Württemburg were also being moved by rail to Brussels, in order to come into the line between Sixth Army and the sea, absorbing III Reserve Corps. Fourth Army's orders were to break through the Allied line from Ypres to the coast and drive on together with Sixth Army to seize Calais.

The Forces Clash at Ypres

The last of Field Marshal French's forces, I Corps under General Sir Douglas Haig, completed its move on 19 October, and French ordered Haig to advance as the spearhead of the BEF, from Ypres through Bruges and on to Ghent, cooperating with the French. After only a few miles the advance turned into an encounter battle as it ran headlong into Fourth Army coming the other way. After three days of fighting both sides took stock, and a small Allied redeployment left Haig with three divisions holding the line east of Ypres, with the French IX Corps to his north and the Cavalry Corps under General Sir Edmund Allenby prolonging the line to the south.

On the German side, Fourth Army reported that it could not break through against the British without more support. A hastily assembled additional force, the equivalent of three corps with extra artillery known after its commander as Army Group Fabeck, was inserted between Fourth Army and Sixth Army to renew the attack. Meanwhile BEF headquarters continued to issue orders for a major advance which Haig increasingly saw as unrealistic. While his troops were of very high qual-ity they were few in number, and they were taking heavy casualties every day. On 30 October he issued orders to dig in and defend.

The critical day of the First Battle of Ypres for both sides came next morning on 31 October. The new German plan was for Army Group Fabeck to make two simultaneous attacks to break the British line and capture Ypres, after which the Germans would sweep onwards to the north and west. One of these two attacks would be made to the south of Ypres, against the British positions on the ridge of high ground which runs southwards from the town and includes the villages of St Eloi, Wytschaete ('Whitesheet' to the British) and Messines. Here the British line was held by dismounted troops of 2nd Cavalry Division and 1st Cavalry Division, with the infantry of 4th Division prolonging the line south of Messines. The cavalry's positions would be subject to a renewed attack by II

Bavarian Corps, which having captured Messines and Wytschaete would swing round to envelop Ypres from the south. Unlike their French and German counterparts, British cavalry were armed with infantry rifles rather than the shorter cavalry carbines, and could hold positions dismounted as well as any infantry. But when dismounted and dug in each cavalry division, even when at full strength, numbered only the equivalent of an infantry brigade.

Rather than following the standard German tactics of a double-encirclement, Army Group Fabeck's second blow would be delivered in the centre by XV Corps and XXVII Reserve Corps, in an attack straight up the gentle slope on either side of the road from Ypres to Menin. Here the key to the British line was the village of Gheluvelt, with its adjacent château and grounds. The British rifle pits and trenches, which lay just in front of the village, were only four miles from Ypres, and were held by Brigadier General C FitzClarence with seven battalions of Major General S H Lomax's 1st Division, with 2nd Division immediately to its north. After several days' hard fighting all the British battalions were seriously understrength. Once Haig's centre had been smashed, XV Corps would continue its advance along the Menin road into Ypres to link up with II Bavarian Corps and complete the British defeat.

This German plan had every chance of succeeding. By hitting the badly stretched British simultaneously in two places, the likelihood was that they would break through to Ypres. With dangerously weak positions and almost no reserves, Haig and Allenby could do little more than plan to hang on to the last man when the expected German attack came, knowing that reinforcements were being scraped together as rapidly as possible. In either sector a German breakthrough resulting in a major British collapse would be absolutely fatal to them. On Messines ridge the British could afford to withdraw a short distance if they did so in a controlled fashion. In the centre at Gheluvelt they could hardly afford to give up a yard. If the British lost Ypres, there were virtually no forces left to defend the English Channel ports, and the way would be open for a major German victory.

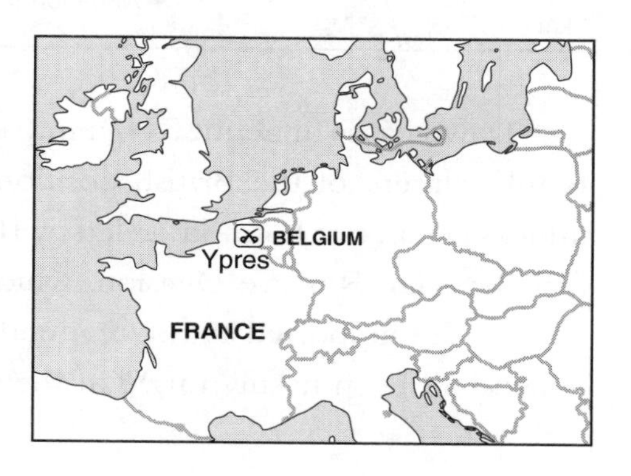

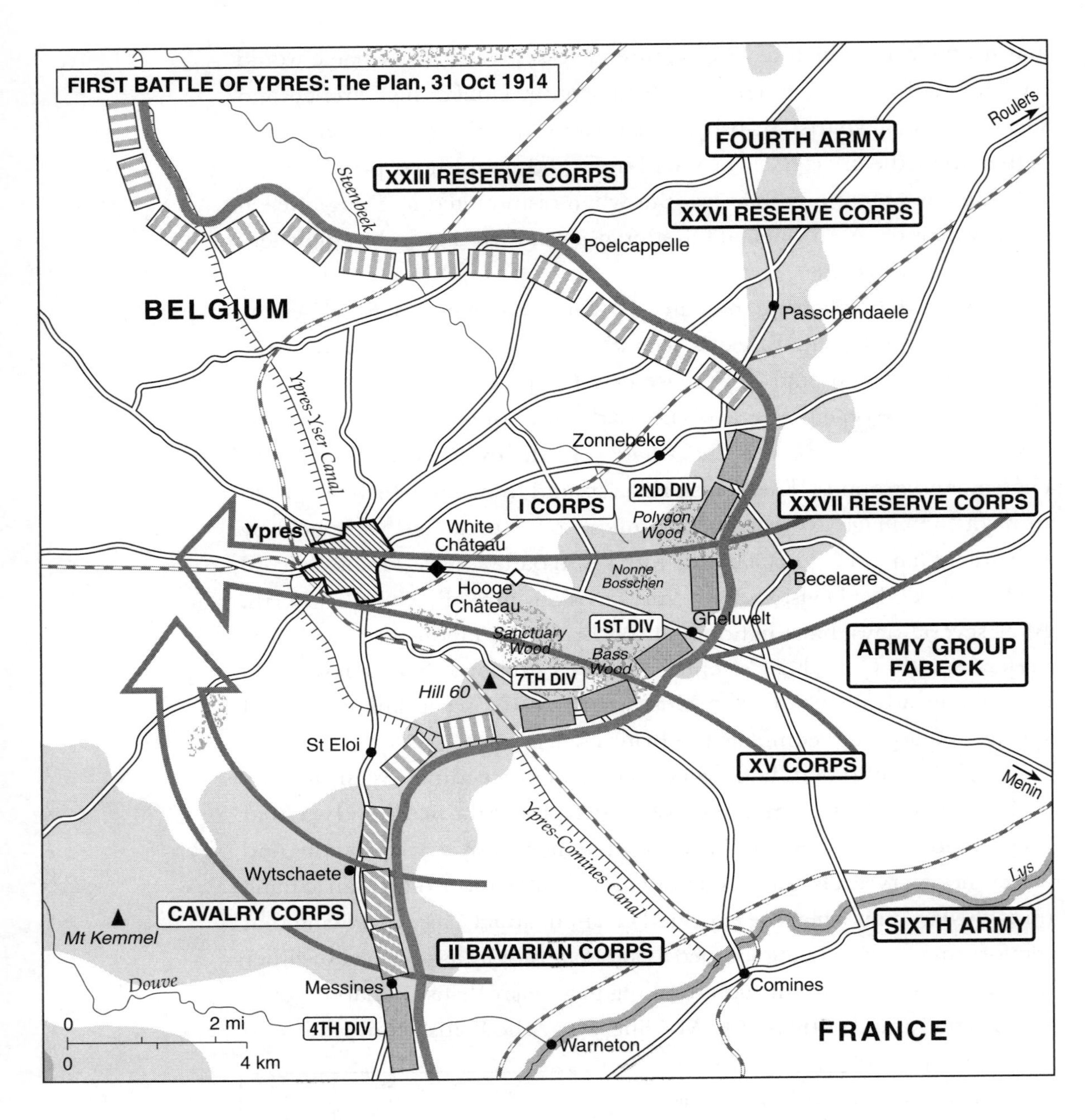

Dawn came at about 6.00 a.m., and with it the first German artillery bombardment of the British positions. On Messines ridge the infantry attacks began two hours later, led by II Bavarian Corps' 26th Division and 3rd Bavarian Reserve Division, which had already seen heavy fighting. These were supported by two regiments of 6th Bavarian Reserve Division, which had been hastily raised at the start of the war. Many of its volun-

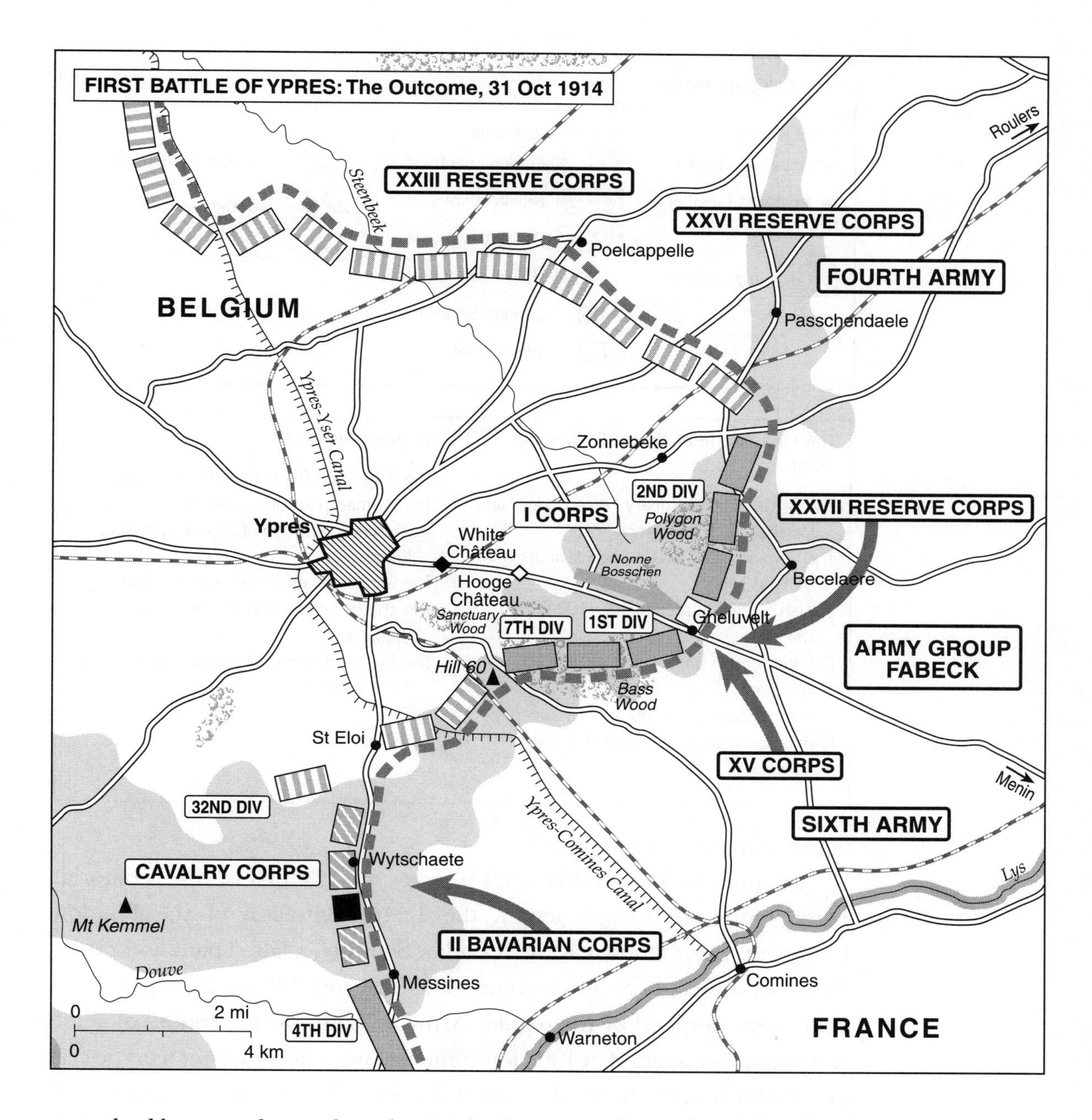

teers had been students when the war broke out, and were later described as going into action with their arms linked together singing Bavarian student drinking songs. But despite the lack of training of some of their formations the Germans had a formidable advantage in numbers. In the centre 2nd Cavalry Division had perhaps 1,500 men in the front line and the same number in reserve, meaning that it was outnumbered by

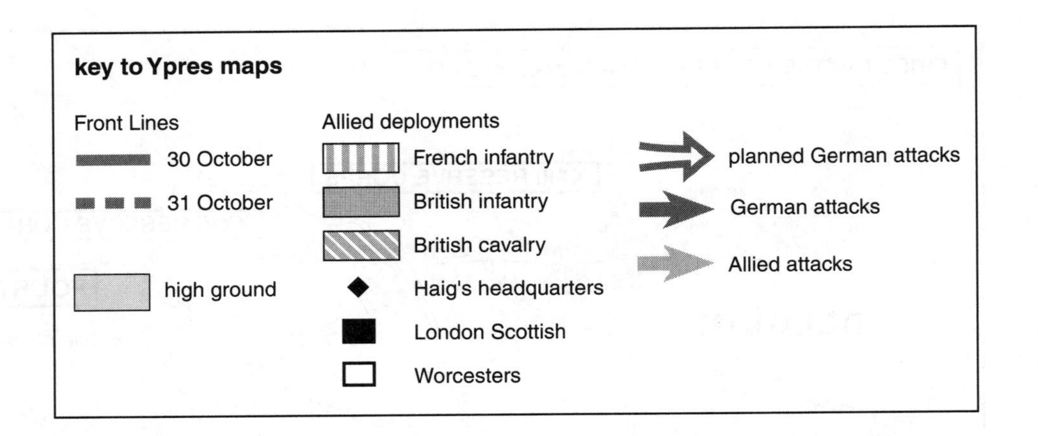

The First Battle of Ypres, 31 October 1914

The British Expeditionary Force under Field Marshal Sir John French
The German Army Group Fabeck under General Fabeck
British forces: approximately 100,000 at the start
German forces: approximately 150,000 at the start
British casualties: 58,000
German casualties: unknown but very heavy

Critical Moments

The Allies and Germans deploy at Ypres
The German plan of attack on 31 October
The London Scottish charge at Messines Ridge
The Germans capture Gheluvelt village
The charge of the 2nd Worcesters at Gheluvelt
The British defence of Nonne Bosschen on 11 November

between five and six to one.

Among the British divisional reserves was the first of the Territorial Force battalions to see action, the 1/14th Battalion of the London Regiment, better known as the 'London Scottish', a kilted battalion raised from Londoners of Scottish origin. The Territorial Force were part-time volunteers, despised by the regular Army as 'Saturday afternoon soldiers', which had been raised in 1908 for home defence only, but which in the crisis of 1914 had been asked to serve overseas. The London Scottish had arrived in France in September for use on lines of communication duties, but in the crisis they had been sent to Ypres, arriving on 30 October and being given first to Haig and then to Allenby.

The battle on Messines ridge was a contest between superior German numbers and weight of artillery on one side, and superior British rifle fire on the other. In the course of the day 26th Division drove the cavalrymen – supported by two British infantry battalions as they arrived – out of

Messines village, fighting house to house. In the centre, 6th Bavarian Reserve Division came forward at the juncture between 1st Cavalry Division and 2nd Cavalry Division, to be met by a bayonet charge by the London Scottish. Digging in on the crest, the Territorials held for the rest of the day against repeated German attacks, the last of which came at about 5.00 p.m. and lasted for an hour. By evening, despite the loss of Messines itself and almost a third of their numbers, the British still held the ridge with more reinforcements arriving, including an infantry brigade from the French 32nd Division.

The British victory had been a narrow one indeed. Of 750 men of the London Scottish, 321 had been killed or wounded. But on the other side the German attack had been shot and fought to a standstill, and the bright young hope of 6th Bavarian Reserve Division had been shattered with hideous losses. To the Germans the attack of 31 October became the *Kindermord von Ypern*, the 'Massacre of the Innocents' at Ypres.

While the contest for Messines ridge was being fought out, the second German thrust was also under way in the centre, aimed at Gheluvelt. An infantry attack coming at dawn was repulsed by the British, except for the loss of an orchard south of the village. This was followed by a heavy artillery bombardment, which reduced houses in Gheluvelt to rubble or set them on fire. Under the weight of this bombardment and renewed German attacks, British casualties mounted to the point at which the battalions of 1st Division almost ceased to exist, being each reduced to fewer than a hundred men commanded by junior officers. On the other side, attacking forces were once more being scraped together from any troops available, including the remaining regi-ment of the novice 6th Bavarian Reserve Division.

The main German attack of the day came at 10.00 a.m. on both sides of the Menin road. Singing and cheering, seven battalions of 30th Division and 54th Württemburg Reserve Division, with six more battalions in reserve, came forward against a British line which by this time numbered no more than a thousand men in total, about the equivalent of one full-strength battalion. But the regular infantry of the BEF were at the time undoubtedly the best in the world for accurate rifle shooting, and they were supported by an equally excellent artillery. Not for the first or last time in the war, the Germans believed that they were facing massed machine guns. Inevitably, weight of numbers and sheer German determination began to tell. Shortly before noon the British battalion defending Gheluvelt village itself was wiped out (only 14 men escaped to safety) and the battered Germans at last took the village. The British line was broken and the way through to Ypres was open.

In the chaos of the battle news took time to travel even the short distance down the Menin road to Haig's I Corps headquarters at the White Château just outside Ypres, a position which would later be known as 'Hellfire Corner'. Staff officers reported the crisis at 1.00 p.m., and after ordering up his reserve, the weak 6th Cavalry Brigade, Haig rode forward on horseback to see for himself. As German fire came closer, troops were fleeing down the Menin road towards Ypres, while artillery was limbering up and the horses pulling the guns to safety. The total collapse of the British line into panic was very close indeed. Haig returned to his headquarters where a last reserve of two Royal Engineer companies were being sent forward to be used as infantry. Orders were drafted – but not yet sent – for a final defensive stand virtually at the foot of the fortified rampart walls of Ypres itself. At about 2.00 p.m. Field Marshal French arrived on foot, having had to walk a mile from his car through the choked and congested roads. He had no help or reinforcements to offer, and the two commanders solemnly said goodbye to each other. Haig then once more mounted his horse to ride up to the front, possibly to die with his men.

Incredibly, on either side of Gheluvelt village the depleted battalions of 1st Division were still clinging on. To the north of the village they held Gheluvelt Château and most of its grounds, while south of the Menin road although another battalion was wiped out by German artillery and infantry attacks (36 men survived) there were no fresh German troops to press the advantage. Those who had captured Gheluvelt, perhaps 1,200 men from 6th Bavarian Reserve Division and 54th Württemburg Reserve Division, were either still engaged fighting the British troops among the trees of the château grounds, or had dispersed in the village in search of water or rest after the morning's battle. The British did not know it, but Gheluvelt village was not properly defended.

Brigadier General FitzClarence had already committed much of his slender reserve to the morning's battles, including one company of the 2nd Worcesters. These troops were part of 2nd Division, but had been placed by its commander at the disposal of 1st Division during the morning, along with three companies from two other battalions. This 2nd Division reserve was now waiting in a wood about a mile to the northwest of Gheluvelt. The plan for how to use the reserve had already been agreed: rather than trying to reinforce the 1st Division line, it would attack directly into the German flank.

Lomax ordered FitzClarence to carry out this plan, and then returned to his own headquarters at Hooge Château, which was shared with 2nd Division headquarters, in order to plan the next move together. As the conference took place German long-range shells hit the château, fatally wounding Lomax and killing and wounding many of the staff officers of both divisions. The commander of 2nd Division was stunned but survived. Knowing nothing of this, at about 1.00 p.m. – just about the time Haig heard of the loss of Gheluvelt – FitzClarence gave the order for the three companies of 2nd Worcesters, 357 officers and men under Major E B Hankey, to attack and 're-establish our line' at Gheluvelt, supported by field artillery fire. The first six hundred yards of the advance was in column of fours through the cover of the trees, with FitzClarence accompanying the battalion. On the edge of the woodland the men halted and shook out into attack formation, with two companies in front and one in support. From this point a further thousand yards of completely open ground stretched out to Gheluvelt, utterly devoid of cover and raked by German shellfire bursting overhead. British stragglers who reached the safety of the treeline cried out that to go forward was certain death. As FitzClarence watched, Hankey gave the order to advance at the double.

Artillery fire cut down over a hundred of the Worcesters before they reached the village and château grounds. But the charge of the remainder drove on through the trees and into the houses, utterly routing the astonished and unprepared Germans and driving them back down the Menin road. In the château grounds the battalion linked up with the British troops still fighting on, and established a line running from the village to the château. FitzClarence sent back word that Gheluvelt had been recaptured and the line restored. The message reached Haig's headquarters just as he was riding out of its gates on his horse. Staff officers recorded their astonishment and jubilation. 1st Division had held and the battle was not lost. An officer was sent after French to give him the good news.

Although the crisis was over, the fighting for the day was by no means finished even at Gheluvelt, but there were no more major German attacks. The total losses for all four companies of the Worcesters that day were 192 officers and men out of 461 all ranks. Next day, German shelling and renewed attacks compelled the British to withdraw from Messines ridge, falling back a few hundred yards to new positions down the slope. They would not retake the ridge until the Battle of Messines in June 1917. The First Battle of Ypres did not end until some days later. On 11 November in miserable weather the Germans renewed their attack on either side of the Menin road with scratch forces, including some battalions of the Prussian Guard. These were met by equally weak and improvised British forces, with a result that was very similar to the fighting at Gheluvelt. This time the wood of Nonne Bosschen a short distance to the northwest, not far from where the Worcesters had waited in reserve on 31 October, was captured. Once more FitzClarence managed to restore the situation, but was himself fatally wounded attempting a final counterattack. This last German failure effectively ended the battle.

The First Battle of Ypres has been called the graveyard of the pre-war British Army. The immense difference between the ambitious plans of both sides and the tiny advances which could be achieved against murderous enemy firepower and tenacity was to make it the forerunner of many Western Front battles. The Allies counted it as a victory for preventing the German breakthrough. But not until October 1918 were they able to advance from Ypres into Belgium in the manner and for the distances that French and Foch had originally planned.

STEPHEN BADSEY

Further Reading

Banks, A., *A Military Atlas of the First World War* (London, 1975)

Blake, R. (ed.), *The Private Papers of Douglas Haig* (London, 1952)

Edmonds, J.E., *Military Operations, France and Belgium, 1914 Volume II* (London, 1933) [British official history]

Terraine, J., *The First World War 1914–1918* (London, 1965)

Crete

20 May–1 June 1941

The Defeat of Both Sides

'The day of the parachutist is over. The parachute arm is a surprise weapon, and without the element of surprise there can be no future for airborne forces.'
Adolf Hitler

Of all battle plans, those involving a radical new departure from conventional thinking carry the greatest risk. This was particularly true of airborne operations in World War II. Parachute and glider troops depended not only for victory but for their very survival on surprise and speed in defeating enemies who were sometimes much stronger, who had reserves much closer, and were not dependent on aircraft with all their associated problems (including the weather) for support. The Battle of Crete, fought almost as an afterthought to the German conquest of the Balkans in spring 1941, ranks as the only complete strategic victory won entirely by airborne forces. But it came very close to being a defeat in battle, and for the German parachute arm it was a defeat in fact.

The Strategic Situation

The island of Crete has always had strategic significance in the Mediterranean. In World War II, it was obvious to all sides that an Allied airfield on Crete could be used to mount raids deep into the Balkans and southern Europe, and as a base for tactical air support of operations in

Greece. In Axis hands it could be used for raids into Egypt and Palestine, and to support operations in North Africa. Moreover, in Souda (Suda) Bay there was a harbour with great potential as a naval base. Obviously, whoever controlled Crete had considerable advantages. Although the British were well aware of Crete's significance, they could take no advantage of it until Greece entered the war in October 1940 when invaded by Italian forces from Albania. At that time there was such a shortage of troops and equipment owing to other British commitments in the Middle East that no more than a small force could be spared to garrison the island.

All this was suddenly changed by the failure of the Allied campaign in Greece in April 1941, and the subsequent evacuation to Crete. About 25,000 men, mainly from the 6th Australian Division and 2nd New Zealand Division, disembarked at Suda Bay, most of them with nothing more than the clothes they stood up in and their personal weapons. General Sir Archibald Wavell, commander in chief in the Middle East, flew to Crete on 30 April to review the situation, and placed Major General Bernard Freyberg, the New Zealand commander, in charge of the defence.

The Rival Plans

Further evacuations gave Freyberg a total of some 30,000 Commonwealth troops and 11,000 Greek troops. On the credit side, some of the evacuation troops had managed to bring heavy equipment, and Freyberg now had a troop of light field artillery. Sufficient machine guns and mortars had arrived to equip all troops, and an additional battalion of infantry was sent from Egypt. Wavell even managed to find a few tanks: 16 Vickers light tanks and 6 Mark II Matilda tanks were shipped to the island, although on arrival they were found to be battered relics of the Desert War, barely able to move.

There was also a handful of Royal Air Force (RAF) fighters on the island. But Freyberg as a realist saw that they were insufficient to have any useful effect and ordered them to be flown to Egypt on 19 May. He then requested permission to spoil the airfields so as to make them unfit for landings. But the RAF, convinced that it would eventually return in strength, was able to prevent this.

The German plan for capturing Crete aimed at gaining command of all the island's strategic points within the first 24 hours. The German force consisted of 22,750 men, 500 transport aircraft, 75 gliders, 280 bombers, 180 fighters, and 40 reconnaissance aircraft. The air assault would be in two waves: at dawn Group West would attack Maleme, and Group Centre would attack Canea (Chania). In the afternoon a second and smaller wave would attack Retimo (Rethymnon) and Heraklion (Irakleion). These various groups were to link up as soon as possible. On the next day, infantry of 5th Mountain Division would be airlifted into the airstrips captured in the initial assault, while the remainder of the division would be shipped across to Heraklion, Souda Bay, and any other available landing place.

The German Airborne Assault

At 7.15 a.m. on 20 May the first German gliders of 1st Assault Regiment landed south and west of Maleme airfield, followed almost immediately by gliders landing in the nearby dry river bed of the River Tavronitis, and on Hill 107 immediately to the south of the airfield. The airfield landing went reasonably smoothly, but the third company, aiming to take the Tavronitis Bridge just west of the airfield, landed in the middle of an area occupied by New Zealand troops, who reacted quickly and inflicted heavy casualties on the invaders. Nevertheless, the assault troops were able to secure the bridge and hold it. While this was in progress, the parachute landings by 7th Paratroop Division began. The 3rd Battalion, intending to drop around Maleme airfield, was dispersed by strong winds and landed among 5th New Zealand Brigade. Within 45 minutes the battalion commander and 400 out of 500 men who had landed were dead, and the battalion was no longer an effective fighting force. Even so, the survivors of 3rd Battalion were due to play a vital part in the battle for Crete.

The attack on Canea was undertaken by two companies landing by glider, while 3rd Parachute Regiment dropped southeast of the town. The gliders which landed northwest of Canea found the area very strongly held, and of 136 men who disembarked from the gliders, 108 became casualties in a very short time. The gliders which landed southeast of the town captured their objectives, but were then unable to link up with the other company owing to the latter's destruction, and withdrew southwards.

By midday things looked black for the Germans. None of their objectives, other than the Tavronitis Bridge, had been taken; casualties had been severe; many commanders were dead, and the various pockets of troops were pinned firmly in place with little prospect of breaking out. But none of this was known at General Kurt Student's headquarters back in Greece, where the second wave was dispatched and arrived over its objectives in Crete at about 3.00 p.m. The 2nd Parachute Regiment dropped at Retimo into an area held by 19th Australian Brigade supported by four battalions of the Greek Army, with the result that the Germans were instantly pinned in their dropping zone. Having landed close to a small hill east of the airfield, the German 1st Battalion managed to collect there and dig in. The 3rd Battalion dropped east of Retimo and also managed to collect themselves, secure a defensive position, and dig in. Between the two battalions the regimental staff also dug in, having dropped with a company of infantry. So the attack at Retimo became a matter of three independent small parties, each pinned down and waiting for either relief or obliteration.

The German 1st Parachute Regiment, meanwhile, had dropped in various localities around Heraklion, and found themselves immediately under fire from British 14th Infantry Brigade which was defending the town. At least one German company was almost entirely killed by the British.

The Lost Opportunity

By the evening of 20 May, the parachute troops were hanging on by the skin of their teeth, scattered all around their objectives, having suffered severe casualties – 1,800 of the 3,000 men who had dropped were dead – and in no shape to take offensive action towards their objectives. By this time, General Student in Greece was beginning to piece together the various reports that were coming to him by radio, and putting the tactic-al picture together. It appeared to Student that Maleme was the key to the battle. If this could be reinforced, then it promised to allow some degree of exploitation. As a result Student gave 5th Mountain Division new orders. Although originally warned to come into action around Heraklion, it was now ordered to land at Maleme, reinforce the paratroops there, and then move east towards its original objective.

That night was in many respects the time when the battle for Crete was won and lost, even though the battle itself continued for some days. Had Freyberg, with his overwhelming superiority in men, artillery and armour, thrown in counterattacks during the night against the exhausted paratroops, he could probably have wiped them out. At least, such attacks would have so disrupted them that they would have been unable to act cohesively on the following day. Moreover, had Freyberg appreciated the significance of the German force around Maleme and re-inforced the airfield, the battle would subsequently have followed a very different course.

This lost opportunity was compounded early next morning, when the commander of 22nd New Zealand Battalion, mistakenly thinking that his forward companies had been overrun, gave orders to pull back. In the confusion the New Zealanders were then genuinely overrun, and Hill 107 was in German hands. This gave them command over the airfield, and with an additional 550 paratroops dropped shortly after this, the airfield itself was captured. Early in the afternoon a solitary Junkers Ju 52 transport aircraft made a successful landing; and even though much of the field was under Allied machine-gun fire, the pilot taxied to the far end of the field from where he was screened from the enemy, unloaded ammunition, took aboard some wounded, and then turned around and made a successful take-off. He immediately reported his success, and the fact that the western end of the airfield lay in dead ground to defensive fire. Within two hours, the first of a string of transports carrying the mountain troops was landing at Maleme. The position was now secure in German hands, and they were not going to be dislodged.

The Battle at Sea

By this time the German seaborne reinforcements had begun their journey. The first group left the Piraeus (the port of Athens) on 19 May, and by the evening of 20 May it had reached the island of Milos, where it anchored for the night. It set sail again next morning, but as it rounded Cape Spatha and came within sight of Maleme the leading flotilla was surprised by a British

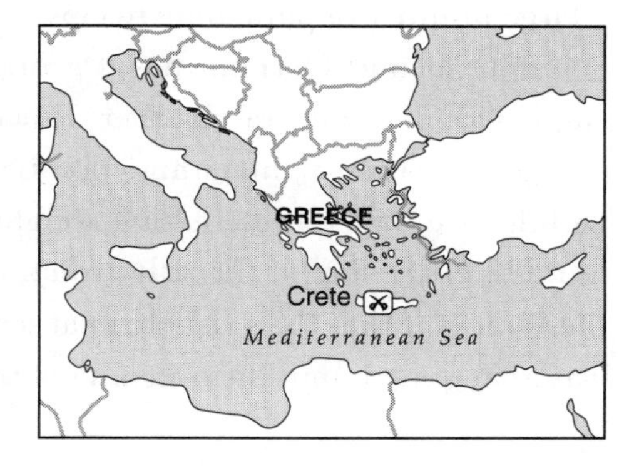

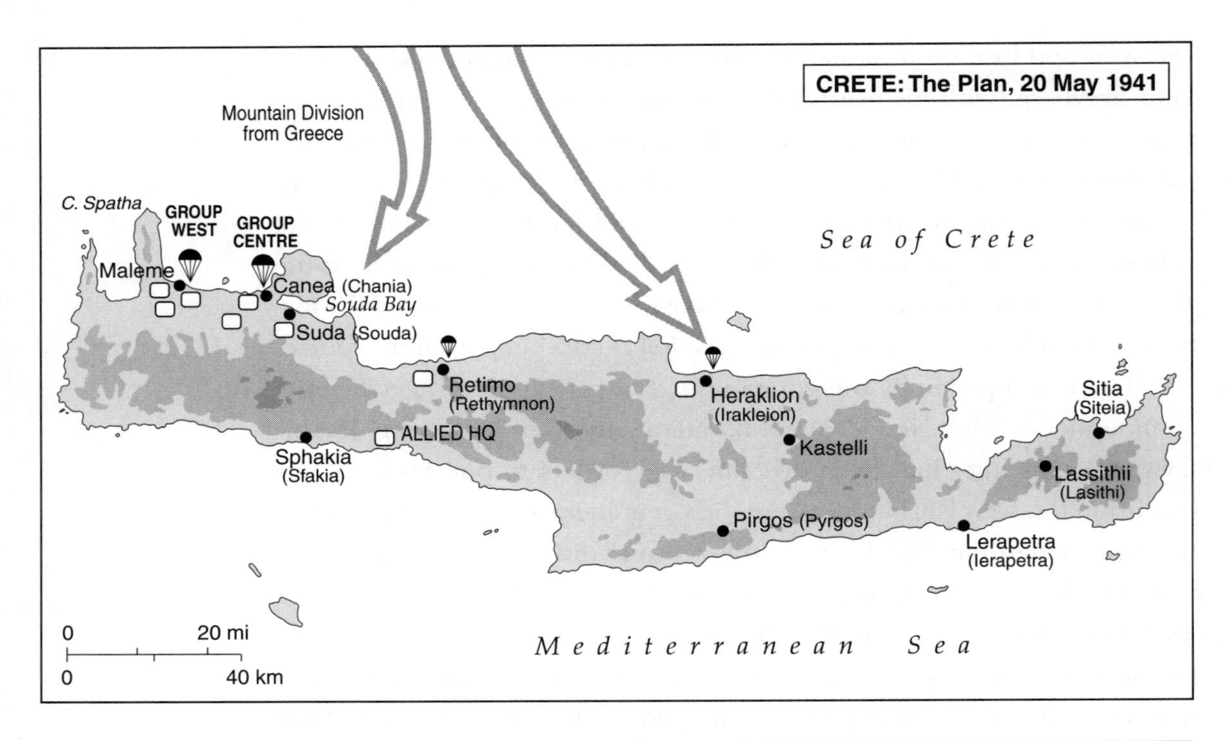

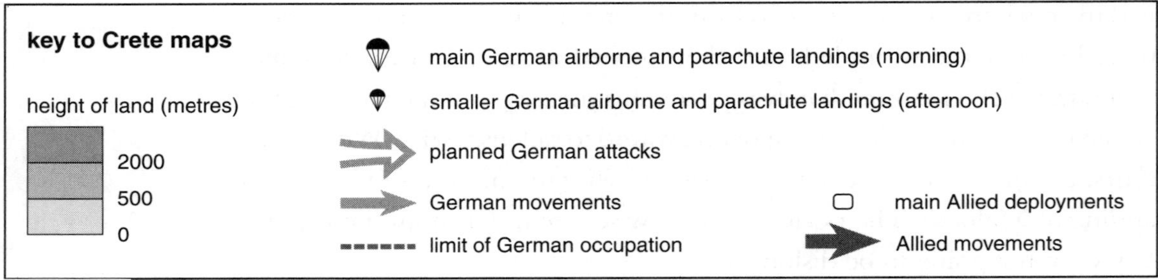

naval force which sank almost every one of its vessels. The 3rd Battalion of 100th Mountain Regiment was destroyed as a fighting force, over 500 officers and men being drowned, together with a large number of parachute troops who were travelling with the flotilla.

The second German naval group left Milos on 22 May, but soon came within range of another Royal Navy force. Before they could engage, German fighters and bombers swarmed over, and the British withdrew under the increasing weight of the German air attacks. Still, mindful of the fate of the first group, the German ships were recalled to the Piraeus rather than risk them at sea, and further reinforcement by sea was abandoned until the outcome of the battle was certain. Meanwhile a

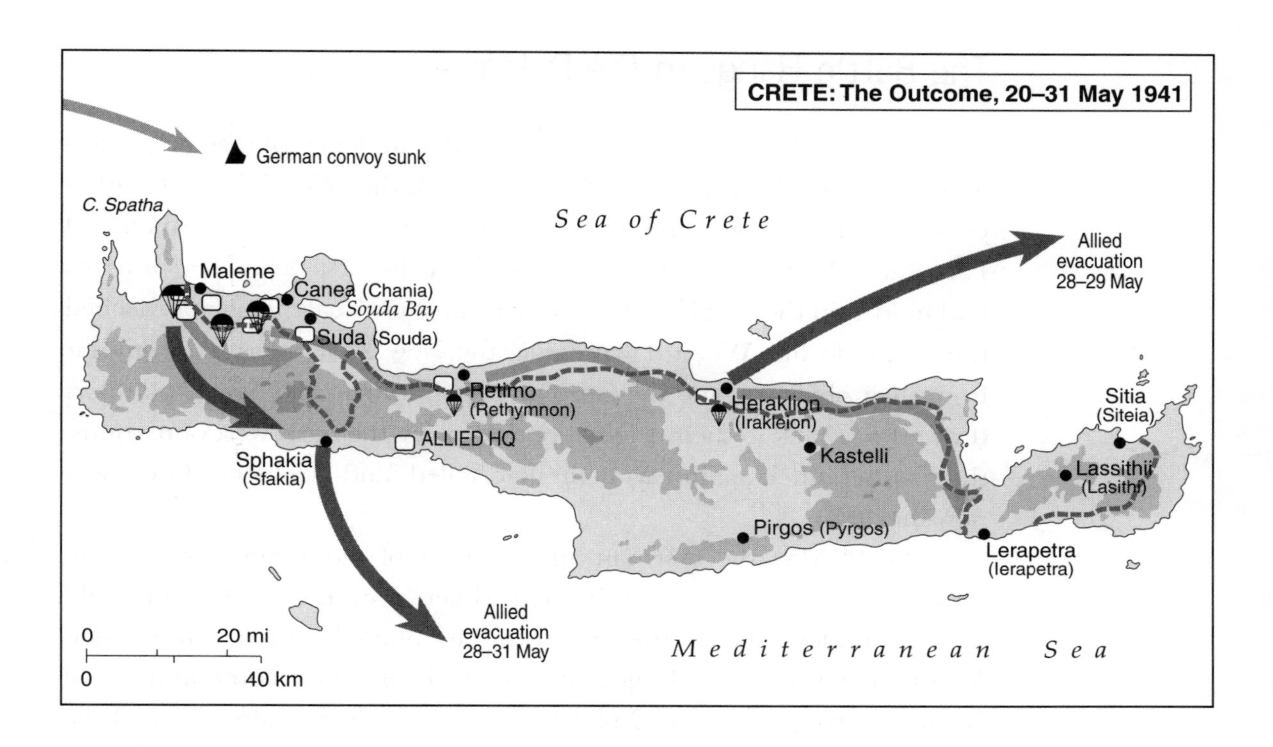

Crete, 20 May–1 June 1941	Critical Moments
Allied forces under Major General Bernard Freyberg	The German dawn airborne landings at Maleme and Canea on 20 May
German forces under General Kurt Student	The afternoon landings at Retimo and Heraklion on 20 May
Allied forces: 44,000 troops	The British intercept the German seaborne convoys on 20 May
German forces: 22,750 troops	5th Mountain Division lands at Maleme on 21 May
Allied casualties: 15,000 (including 11,835 prisoners)	The German attack at Galatas on 25 May
German casualties: 9,000 (including 6,000 dead)	The Allies withdraw southwards and evacuate Crete

major air assault by the Luftwaffe (German airforce) was mounted against any British ships that could be found. Very quickly the cruisers *HMS Gloucester* and *Fiji*, and the destroyers *Juno* and *Greyhound* were sunk, while the cruisers *Naiad* and *Carlisle*, and the battleships *Warspite* and *Valiant* were severely damaged. It became obvious that British naval forces could no longer operate within flying range of the German airbases during daylight.

The Battle Hangs in the Balance

During the night of 21–22 May, the New Zealanders mounted a counterattack to retake Maleme airfield. But to reach the field, the attack had to cross ground that was infested with the remnants of the German 3rd Parachute Battalion, virtually destroyed on the first day. The survivors had hidden in the rough ground, and as the New Zealanders approached they came to life. When dawn came, the New Zealanders were still far from their objective, and the German dive-bombers and fighter-bombers drove them back to their original positions. By the narrowest of margins, the counterattack against Maleme had failed, and the loss of Crete was now inevitable.

On 22 May the German air transport of mountain troops to reinforce the island continued through Maleme, even though the airfield was still under artillery fire. The 5th Mountain Division commander, Major General Julius Ringel, also arrived and took command of all German troops in the Máleme area, dividing them into three *Kampfgruppe* ('battle groups'). One of these was to defend the Maleme area from attacks and expand westward to capture nearby Kastélli; the second was to move north to the sea, and then begin extending eastward along the coast; while the third was to move eastward across the mountains to outflank the Allied positions.

These moves began at dawn on 23 May, and were generally successful. One result was that the New Zealanders were forced to pull back as the mountain troops outflanked them, and this led to the withdrawal of their covering artillery to a more secure position, so that the Maleme airfield was no longer within their range. The Germans could now fly in without fear of interruption.

The Decisive German Attacks

At noon on 25 May the decisive battle for Crete began with a concerted German attack against the blocking position at Galatás, southwest of Canea on the road from Maleme. After bitter fighting the New Zealanders were eventually ousted from the village and the Germans moved in, only to be thrown out by a New Zealand counterattack led by two tanks. But by now the New Zealand numbers were too depleted, and during the night they withdrew, so that at dawn the German mountain troops were able to move back in and secure the village, so opening the road for their advance to Canea.

On the evening of 26 May Freyberg signalled to Wavell, '... in my opinion the limits of endurance have been reached by the troops under my command... from a military point of view our position here is hopeless... Provided a decision is reached at once, a certain proportion of the force might be embarked.' The Allies concluded that they had lost the battle. On 27 May the decision to evacuate Crete was taken, and the defenders began to withdraw southward. The Germans, themselves stretched to the limit, failed to realize what was happening and continued their attacks against Canea, while to the southwest of Suda their attention was fully occupied by equally violent Allied counterattacks. The Germans failed to see this sudden activity for what it was, a rearguard action intended to cover the withdrawal of the remainder of the Allied forces.

Late on 27 May, Major General Ringel formed another Kampfgruppe, and sent it along the coast road to relieve the hard-pressed paratroops still holding out in their scattered positions around Heraklion and Retimo, with orders to combine with them and capture Heraklion airfield. The Germans moved off early on 28 May, but were stopped only three miles beyond Souda by a party of British Commandos from 'Layforce' (named after their commander, Colonel R E Laycock), who had been landed at Suda during the last few nights. Together with two battalions of New Zealanders, Layforce held up the German advance for several hours. But eventually the Allies were beaten back, and the Kampfgruppe was able to make contact with other Germans still fighting around Stilos, about five miles southeast of Souda on the Retimo road.

As the New Zealand and Australian troops withdrew, the Germans were able to restart their eastward march. But after a few miles they were again embroiled in bitter fighting when they ran into the main body of Layforce, accompanied by some Australian infantry. Eventually one German group, using its mountain skills, was able to outflank the fighting area and turn the Allied position. During the night Layforce withdrew once more, and in the morning the German eastward advance resumed. The Germans finally reached Retimo at about 1.00 p.m. on 29 May, making contact with the survivors of the paratroops still there.

The Allied Evacuation

Following Freyberg's orders on 27 May to evacuate Crete, the garrison at Heraklion embarked during the night of 28–29 May aboard two Royal Navy cruisers and six destroyers. On 29 May German troops approached Heraklion and, finding no resistance, took the town. The Germans had made a simple error which turned out to be the salvation of many Allied soldiers: they assumed that the Allies were retreating eastward, when the main retreat was southwards towards the little port of Sfakia, from which the Royal Navy was evacuating troops as fast as it could. The Germans realized their mistake on reaching Ierapetra (Lerapetra) in the southeastern corner of the island. It was obvious that the Allies could not be retreating eastwards, and on 29 May the Germans resumed their advance to the south.

By the evening of 30 May the leading German elements were within three miles of Sfakia, and the rest of the island was in German hands. By then, close to 12,000 Allied troops had been removed by the Royal Navy from Sfakiá, and the evacuation continued in the face of severe German air attacks. Major General Freyberg left by flying boat on 30 May, leaving Major General E C Weston in command of the rearguard. Weston was then also ordered out, handing over to Laycock, with orders that the remaining Allied troops should surrender at 9.00 a.m. on 1 June. By 4.00 p.m. on that day, German forces were in complete control of the island, taking prisoner almost all the remaining Allied troops, about half of Freyberg's command.

Crete was a German victory, but a very costly one. On the Allied side, the Royal Navy suffered severe damage to one aircraft carrier, three battleships, six cruisers, and nine destroyers, and lost three cruisers and six destroyers sunk with the loss of over 2,000 sailors. About 3,500 Allied troops had also been killed or wounded. But of the 22,000 German troops who had arrived on the island some 6,000 had been killed. The most significant result of the Battle of Crete was the conclusion drawn by Adolf Hitler from the fearful German casualty list. The Germans would never again mount a major airborne operation.

IAN HOGG

Further Reading

Beevor, A., *Crete, the Battle and the Resistance* (London, 1992)
Buckley, C., *Greece and Crete 1941* (London, 1977)
Liddell Hart, B. H., *The Other Side of the Hill* (London, 1948)
Lucas, J., *Storming Eagles* (London, 1988)

Index